ASAE Publication 23-81

IRRIGATION SCHEDULING FOR WATER & ENERGY CONSERVATION IN THE 80'S

The Proceedings of the
Irrigation Scheduling Conference

December 14-15, 1981
The Palmer House
Chicago, Illinois

Published by
American Society of Agricultural Engineers
P.O. Box 410 • St. Joseph, Michigan 49085

Library of Congress Card Number (LCCN) 81-70534
International Standard Book Number (ISBN) 0-916150-42-9

TABLE OF CONTENTS

* Abstract of Alternate paper
† Alternate paper
‡ Abstract only

CONFERENCE ON IRRIGATION SCHEDULING
FOR WATER & ENERGY CONSERVATION IN THE 80's

PREFACE

The Irrigation Management Committee of the American Society of Agricultural Engineers decided, in 1979, that the time was appropriate to establish the state of the art in irrigation scheduling. Planning began for a conference to highlight recent scheduling advances for the benefit of policy makers, consultants, engineers, technicians and others. This subject matter related closely to the theme of the ASAE Annual Winter Meeting that immediately followed the conference, "Land and Water Resources - A Critical Responsibility". The organizations listed on the following page accepted invitations to cooperate in sponsorship of the conference.

Papers reporting research results, design and analysis techniques, theory, further research and development needs and applications were solicited. Nearly one hundred paper proposals were received from which 26 papers and four alternates were selected. From these, three one-half day sessions were developed, featuring the following main topic areas:

1. 1st Session - Algorithms and software; ET and crop growth models; verification of procedures; theoretical aspects; water flow modeling; problems and research needs; etc...

2. 2nd Session - Methods, equipment, system design, communication systems, problems and technology needs, etc...

3. 3rd Session - Commercial applications, consultant and technical services, application problems, education and extension.

The final presentation of the Conference, by M.E. Jensen, National Program Leader, USDA, ARS, summarized current irrigation scheduling technology and challenged the engineering profession to extend research and application of recently developed knowledge.

The Conference schedule was arranged to encourage discussion by the audience. Because these Proceedings were printed before the Conference, the discussions are not included herein.

Cooperating societies and members of the various sub-committees involved in the Conference planning are listed on the following page. Their active participation has contributed greatly to the quality of this Conference.

The Conference committee would also like to express its appreciation to the authors for their contributions and for working under a rigorous schedule to ensure that these Proceedings were ready on time.

We hope that this Conference and its Proceedings will be of value to you in increasing or updating your knowledge of irrigation scheduling technology.

C.J. Phene
Chairman

E.C. Stegman
Vice-Chairman

SPONSORED BY

The American Society of Agricultural Engineers
P.O. Box 410, St. Joseph, MI. 49085 (616) 429-0300

COOPERATING ORGANIZATIONS

American Society of Agronomy
Soil Science Society of America
American Society of Civil Engineers
Irrigation Association
U.S. Committee of the International Commission on Irrigation, Drainage and Flood Control

CONFERENCE PLANNING COMMITTEE

C.J. Phene, Chairman
E.C. Stegman, Vice-Chairman
J.A. Wright, Secretary
M.A. Purschwitz, ASAE Staff
T.A. Howell, Program Co-Chairman
J.R. Lambert, Program Co-Chairman
E.G. Kruse, Publication Chairman and Program Committee
A.G. Smajstrla, Publicity Chairman
D.H. Hay, Finance Chairman
L.F. Hermsmeier, Local Arrangements Chairman
D.F. Heermann, Program and Publication Committee
F.H. Siddoway, Program Committee
J.L. Wright, Publication Committee
C.L. Anderson, ASAE Irrigation Group Chairman
J.R. Gilley, ASAE Irrigation Group Vice-Chairman

The following represented cooperating organizations:

E.G. Kruse (ASCE)
L.N. Mielke (ASA/SSSA)
T. Schiltz (IA)
J Lord (ICID)
J Buchheim (ICID)
J.M. Lima (Industry)

THERMAL INFRARED TECHNIQUES FOR ASSESSING PLANT WATER STRESS

P. J. Pinter, Jr. R. J. Reginato

A quantitative estimate of impending plant water stress is a critical prerequisite for the efficient scheduling of irrigations. Ideally, the warning of stress should come from the crops themselves, obviating the need for precise information regarding available soil moisture, root distribution and evaporative demand of the atmosphere. Unfortunately, many of the time honored and accepted methods for quantifying stress by means of physiological plant measures are both labor intensive and tedious. Furthermore, they are subject to considerable experimental and sampling error (Ritchie and Hinckley 1975). In fact, the variability between plants is often so large that it obscures any real differences which might otherwise be attributed to plant water status. Physiological approaches often involve measurements on single plants, leaves or petioles; as such they are point measurements, many of which may be required to characterize a single field because of differences in soil properties. Examples include xylem pressure potential (Scholander et al. 1965), leaf diffusion resistance (Kanemasu 1975) and petiole and leaf water contents (Longenecker and Lyerly 1969). Although various indicators of this type have been suggested to schedule irrigations to achieve increased water use efficiency (Namken 1965, Hiler and Clark 1971, Hiler et al. 1974, Misra and Pant 1981), economic considerations will probably limit their widespread acceptance (Stegman et al. 1976).

The relationship between plant leaf temperatures and moisture stress has been qualitatively documented for a number of years (Wiegand and Namken 1966, Ehrler 1972, Jackson in press), and evidence is now accumulating which establishes infrared (IR) thermometry as a reliable surrogate for certain physiologically-based water stress measurements. Ehrler et al. (1978) showed an inverse relationship between the stress degree day parameter [i.e., SDD = canopy temperature (T_c) minus air temperature (T_a)] and the xylem pressure potential (ψ_ℓ) of wheat. Later, Idso et al. (1981a) refined the SDD by taking the evaporative demand of the atmosphere into account. The transformed stress parameter which is termed the crop water stress index (CWSI), was well correlated with ψ_ℓ in alfalfa plants subjected to varying degrees of water stress (Idso et al. 1981b). The major advantage which IR thermometry offers over conventional stress assessment techniques is the ease and rapidity with which plant temperature measurements can be made (Pinter in press, Jackson in press). Since entire fields can be surveyed in a very short period of time, this technique appears to offer enormous potential for scheduling irrigations on a cost effective basis (Geiser et al. 1981, Jackson et al. 1980).

The objectives of the research reported here are threefold: first, to establish a relationship between the CWSI and the ψ_ℓ of cotton leaves; second, to examine this relationship throughout the season for differentially-irrigated cotton plots and; third to determine its potential utility for scheduling irrigations.

The authors are: P. J. Pinter, Jr., Research Entomologist and R. J. Reginato, Soil Scientist, USDA, ARS, U. S. Water Conservation Laboratory, 4331 E. Broadway Road, Phoenix, AZ 85040.

EXPERIMENTAL METHODS

The cotton experiment[1/] was conducted during 1980 in a 2-ha field located on the University of Arizona, Cotton Research Center Farm in Phoenix, AZ. The soil is an Avondale Loam [a fine loamy, mixed (calcareous), hyperthermic, Anthropic Torrifluvent]. Short staple cotton (Gossypium hirsutum L. var. 'Deltapine-70') was planted in E-W oriented rows on 14 April 1980 and then thinned to a stand density of 86,000 plants/ha. Briefly the experiment consisted of six replications of six different early season irrigation treatments. The final irrigation was applied in mid-August. Only two replications were used for the data reported here.

Cotton leaf xylem pressure potentials were measured with a pressure bomb apparatus using techniques similar to those proposed by Scholander et al. (1965). Measurements were taken at irregular intervals during the season, but mainly on the day before an irrigation, then once again 2-3 days following irrigation. Three sunlit leaves (at the 4th node from the top) were selected and cut from each field plot. They were immediately placed in plastic bags and stored in a moist insulated chest: measurements were usually made within 45 minutes of field collection. Determinations were made from 1200 to 1500h (MST), an interval during which cotton ψ_ℓ remains relatively constant (R. Reginato and S. Farah, unpublished data).

Radiant leaf and canopy temperatures were obtained between 1330 and 1345h (MST) using a portable handheld infrared thermometer (10.5 to 12.5 μm band-pass filter, 4°FOV) that was calibrated for use in high ambient air temperatures. Average leaf temperatures (T_ℓ) were obtained by aiming the IR thermometer at 8 individual expanded leaves selected at random from the top of the canopy. Average canopy temperatures (T_c) were taken with the radiometer pointed obliquely towards the crop (about 15-25 degrees from the horizontal) and at right angles to the row direction (4 viewing north and 4 viewing south). It is important to make the distinction between T_ℓ and T_c. The former represents temperatures of plant parts only while the latter may include a composite of plant and background soil temperatures.

No corrections were made in the apparent temperatures for leaf or canopy emissivities; nor was the radiant flux from the sky considered in making the temperature observations. Neglecting these parameters introduces a variable 1 to 1.5°C negative bias in the absolute temperatures, however, they can be ignored when the primary concern is the relative differences which exist between treatments or day to day changes within a single treatment (Perrier 1971). At the start and finish of each set of IR measurements, wet and dry bulb temperature measurements were made with an aspirated psychrometer held at a height of 150 cm above the soil. Solar radiation was measured with a hemispherical pyranometer.

RESULTS AND DISCUSSION

Calculation of the CWSI

To determine whether IR thermometry would provide a reliable method for following changes in cotton plant water status from one irrigation to the next, the T_ℓ data were transformed into the CWSI parameter proposed by Idso et al. (1981a). First, it was necessary to establish expected leaf temperatures for non-stressed plants as a function of air temperature (T_a, °C)

1/ The experiment was designed and maintained by Drs. K. E. Fry, G. Guinn, and J. R. Mauney, plant physiologists at the USDA, ARS Western Cotton Research Laboratory, 4135 E. Broadway Rd., Phoenix, AZ. K. E. Fry is presently at the Cotton Research Center, 4207 E. Broadway Rd., Phoenix, AZ.

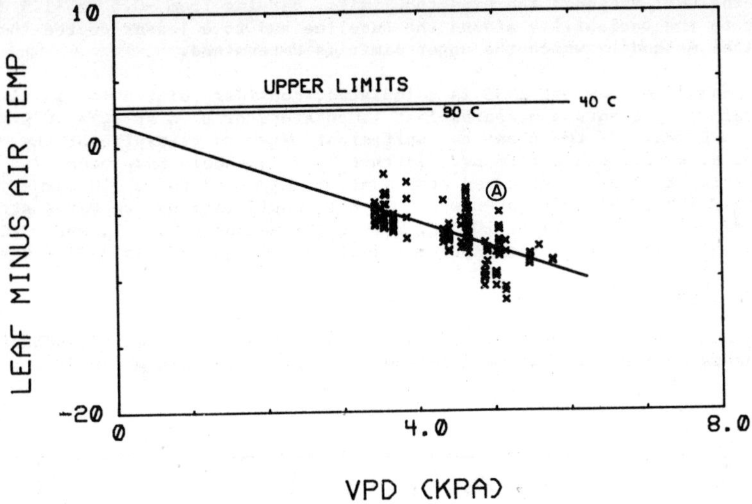

Fig. 1. T_ℓ minus T_a as a function of vapor pressure deficit for non-stressed cotton plants. Horizontal lines show theoretical upper limits for plants which are stressed so severely that transpiration has ceased.

and vapor pressure deficit (VPD, kPa). Accordingly, we selected a subset of T_ℓ data consisting of temperatures obtained on clear days when the field plots were assumed to be non-stressed. Then, the differences between T_ℓ and T_a (i.e., the SDD) were graphed as a function of VPD (Fig. 1). The resulting baseline was described by the linear relation

$$SDD = 1.71 - 1.90\ (VPD) \qquad \text{EQN 1}$$

$$r = 0.67 \qquad n=133$$

The data show that non-stressed cotton maintains apparent leaf temperatures an average of 5 to 10°C cooler than air temperatures under ambient vapor pressure deficits normally encountered during Arizona summer conditions. When water in the effective rooting zone of the crop becomes limiting, leaf temperatures rise above this baseline because transpirational cooling is reduced. In fact, we interpret the scatter of data points about the baseline in Fig. 1 as an indication that at times the plants in certain plots might not have been transpiring at a potential rate. This does not, however, diminish the utility of the baseline as a reference point for comparison purposes. When transpiration ceases altogether, $T_\ell - T_a$ attains an upper limit which is specified by radiant and convective energy exchanges. Idso et al. (1981a) indicated that the upper limit varies with ambient temperatures, and that for practical purposes it can be approximated by an empirically-based algorithm which takes the vapor pressure gradient between T_ℓ and T_a (at VPD = 0) into account. As an example, upper limits for a severely stressed, non-transpiring crop are shown on Fig. 1 for T_a of 30 and 40°C.

The CWSI is defined within the context of the range over which T_ℓ can vary due to water stress conditions. It is the ratio of the deviation of the measured $T_\ell - T_a$ from the well watered baseline to the complete range for a given VPD. The CWSI is dimensionless and in theory should progress from 0 for non-stressed plants transpiring at potential rates to 1 for severely stressed plants which are not transpiring. In this experiment with cotton we found the CWSI exceeded the expected limits, varying from -0.5 to +1.5 due mainly to the variability around the baseline and to a lesser degree the empirical method by which the upper limit is determined.

As an example of how the CWSI is calculated, consider point A in Fig. 1. This value represents a measured leaf temperature of 37°C at T_a = 40°C and VPD of 5.0 kPa. If the plant had sufficient water to transpire at the potential rate, we calculate from Eq. (1) that $T_\ell - T_a$ should have been -7.8°C. Conversely, a severely stressed crop would be expected to be 3°C warmer than the air. The total range over which $T_\ell - T_a$ could vary due to water stress would be +3°C - (-7.8°C) or 10.8°C. Since the measured $T_\ell - T_a$ was -3°C, the plants were 4.8°C above the expected baseline and the CWSI is 4.8/10.8 = .44.

Trends in CWSI

The season long trend of CWSI (based on T_ℓ) is shown for two representative treatments in Fig. 2. The CWSI follows a cyclical pattern which is

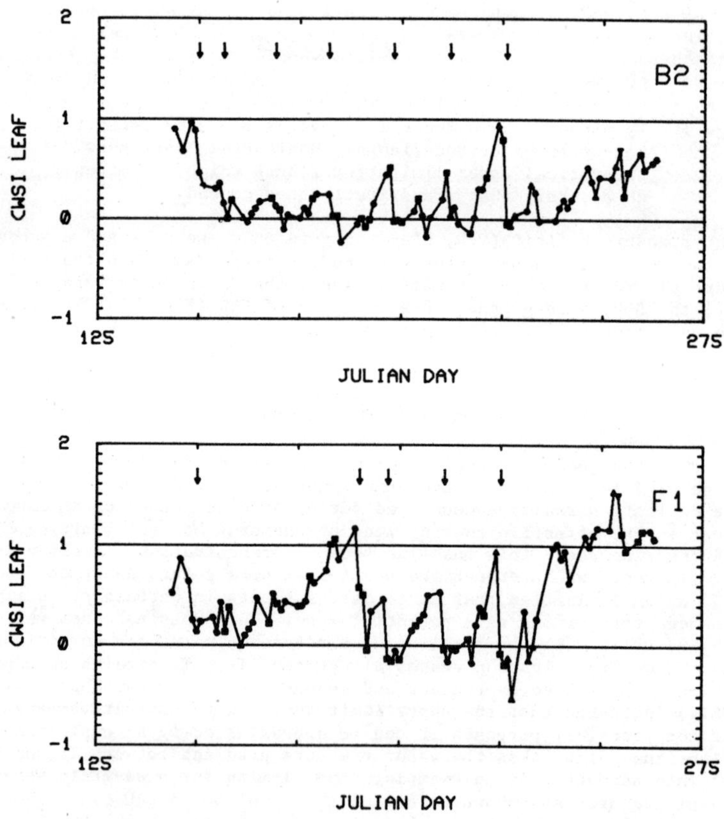

Fig. 2. Seasonal trend of $CWSI_{leaf}$ for two early season cotton irrigation treatments. Arrows along upper axis represent irrigation events.

synchronous with the irrigation events (indicated by arrows along the upper margins). Almost immediately following an irrigation the CWSI falls to a minimum value and then climbs slowly as the crop exhausts its supply of water. The rate of increase in CWSI between irrigations appears directly related to the atmospheric demand and the size of the plants and inversely related to the availability of water stored in the soil. Note that the CWSI does not return to zero following irrigations early in the season. It appears that leaf temperatures of smaller plants are more closely coupled to their thermal radiant environment than those of larger plants. As a consequence, it may be that non-stressed plants will require a different baseline early in the season.

Comparison of CWSI with Leaf Xylem Potential

The ψ_ℓ data for all treatments were paired with corresponding daily values of CWSI and analyzed by simple polynomial regression techniques. Residual analysis indicated that although these two variables were very well correlated from mid-June until mid-August, prior to that time the CWSI predicted a more negative ψ_ℓ than was found. After mid-August the reverse was true. To determine which additional environmental and crop factors were most important in determining ψ_ℓ and to compensate for this seasonal effect, stepwise multiple linear regression analysis was employed (Table 1). Leaf xylem pressure potential was designated as the dependent variable; the independent variables were the age of the crop, air temperature at 150 cm, vapor pressure deficit, solar radiation, and CWSI based on both T_ℓ and T_c.

Table 1: Variables Tested in the Stepwise Multiple Regression Procedure.

Variable	Symbol	Status	Units	Range	Mean	SD
Leaf xylem pressure potential	ψ_ℓ	dependent	MPascals	-1.58 to -3.55	-2.54	0.54
Age of crop	Age	independent	days from planting	43 to 132	111	31.9
Air temperature (+150 cm)	T_a	"	°C	29.5 to 40.9	36.0	2.83
Vapor pressure Deficit (+150 cm)	VPD	"	kPascals	2.11 to 6.71	4.48	1.06
Solar Radiation	I	"	W m^{-2}	757 to 999	886	46.1
CWSI leaf	$CWSI_\ell$	"	dimensionless	-0.33 to +1.15	+0.37	0.34
CWSI canopy	$CWSI_c$	"	dimensionless	-0.50 to 3.05	+0.52	0.62

Correlations Between Variables

Simple partial correlations between the variables are summarized in Table 2. Leaf xylem pressure potential was most highly correlated with CWSI leaf and also with the age of the crop. Note that the correlation between CWSI canopy and ψ_ℓ was not highly significant. This was due to the influence of varying amounts of soil viewed by the IR thermometer early in the season when the plants were small and also later in the season when lodging opened several of the canopies. A significant partial correlation between variables does not necessarily imply a cause-effect relationship (viz. solar radiation and crop age), however, the absence of significance suggests that the variables are independent of one another. An important finding was that CWSI leaf was not correlated with air temperature, vapor pressure deficit or solar radiation. Leaf xylem pressure potential also appeared independent of VPD if only partial correlations are considered. But as will be shown in the next section, VPD does exert a second order modifying effect on ψ_ℓ once the primary effects of water stress have been removed.

Table 2: Matrix of partial correlations between variables listed in Table 1 (n = 147).

	ψ_ℓ	Age	T_a	VPD	I	$CWSI_\ell$	$CWSI_c$
ψ_ℓ	1.000	-0.548**	-0.135	-0.139	-0.203*	-0.672**	-0.176*
Age		1.000	-0.220**	-0.387**	-0.673**	-0.176*	-0.419**
T_a			1.000	0.868**	0.590**	0.090	0.188*
VPD				1.000	0.630**	0.003	0.145
I					1.000	0.045	0.174*
$CWSI_\ell$						1.000	0.568**
$CWSI_c$							1.000

* Significant partial correlation at 0.05 >P> 0.01.
** Significant partial correlation at P < .01.

Multiple Linear Regression

The stepwise regression procedure established that the CWSI based on T_ℓ explained 45% of the variability in ψ_ℓ (Table 3). Subsequent inclusion of the age of the crop and VPD improved the prediction of ψ_ℓ significantly. It is believed that the age of the crop is probably a surrogate for a physiological or morphological plant response which was not measured (i.e., osmotic pressure potential, leaf thickness etc.). The fact that VPD improved the estimate was not totally unexpected. Even though the CWSI is normalized to account for VPD, it probably also exerts a direct effect on ψ_ℓ, as shown for alfalfa by Idso et al. (1981b). The final equation for predicting ψ_ℓ of cotton was

$$\psi_\ell = -0.274 - 0.905(\text{CWSI}) - 0.010(\text{age}) - 0.19(\text{VPD}) \quad \text{EQN 2}$$

$$r = 0.87 \qquad n = 147$$

where ψ_ℓ is in MPascals, CWSI is based on apparent T_ℓ, the age of the crop is in days from planting and VPD is in kPascals. The standard error for the prediction of ψ_ℓ from the above independent variables was 0.269 MPa. The regression analysis showed that the addition of air temperatures, solar radiation or CWSI canopy as independent variables did not improve the estimate significantly.

Table 3: Summary of stepwise multiple linear regression relating the xylem potential (ψ_ℓ) of cotton leaves at 1200-1500 h (MST) to plant and environmental parameters. Data were based on 147 observations of mean ψ_ℓ throughout the season.

Step	Added Variable	Predictive Equation	r	SE of Predicted ψ_ℓ (MPa)
1	$CWSI_{leaf}$	ψ_ℓ = -2.144 - 1.073(CWSI)	0.67	0.398
2	Age of Crop	ψ_ℓ = -1.364 - 0.948(CWSI) - 0.007(Age)	0.80	0.323
3	VPD (kPa)	ψ_ℓ = -0.274 - 0.905(CWSI) - 0.010(Age) -0.19(VPD)	0.87	0.269

Trends in Predicted ψ_ℓ

Equation 2 was used in a predictive mode to calculate the ψ_ℓ throughout the season for all days when the CWSI measurements were obtained. Figure 3 shows the results for the same two treatments portrayed earlier and permits the data to be visualized in a more conventional stress format for the entire season, even though actual ψ_ℓ measurements were only available for a relatively few days for each treatment. Actual ψ_ℓ measurements are shown as circles while the predicted values are shown as a continuous line in Fig. 3.

Good correspondence between the actual and predicted data points was expected, since the former were used to generate the model. The important concept which Fig. 3 illustrates is that ψ_ℓ values decrease in a regular fashion between irrigations. Xylem pressure potential attains its least negative value immediately following an irrigation. Then as the plants extract increasing amounts of water from the soil, the ψ_ℓ declines to minimum levels. Some day to day variability in the data still remain but the trends between irrigations are evident and can be projected to permit advance decisions for timing the next irrigation. If one were to establish a minimum acceptable level, say -2.5 MPa, for signaling the need for irrigation, then use of Eqn. 2 and IR thermometry could provide a viable means to attain that goal without resorting to the tedious and time consuming pressure bomb technique. In the case of plot F2 (bottom of Fig. 3) it is clear that the second irrigation was delayed far beyond a ψ_ℓ which might have signalled that the plant was severely stressed. However, we wish to point out that a given value of ψ_ℓ may not be the best criterion on which to base irrigation scheduling decisions. This conclusion arises from the observation that ψ_ℓ is dependent on ambient VPD. Equation 2 shows that ψ_ℓ decreases 0.2 MPa for a 1 kPa increase in VPD. Idso et al (1981b) have shown ψ_ℓ changes of a similar magnitude for well-watered alfalfa and wheat, and more recently for cotton also (Idso et al unpublished data). For ψ_ℓ to be of greatest utility in scheduling irrigations it must be normalized for VPD.

In summary, we have shown that the CWSI, the age of the crop and VPD can be transformed into a format which is more readily interpreted from a crop water stress standpoint. Although criteria for irrigating cotton in these experiments were based on preselected calendar dates, it is not difficult to visualize how predicted ψ_ℓ data, generated from easily acquired radiant leaf temperatures, could be used to schedule irrigations.

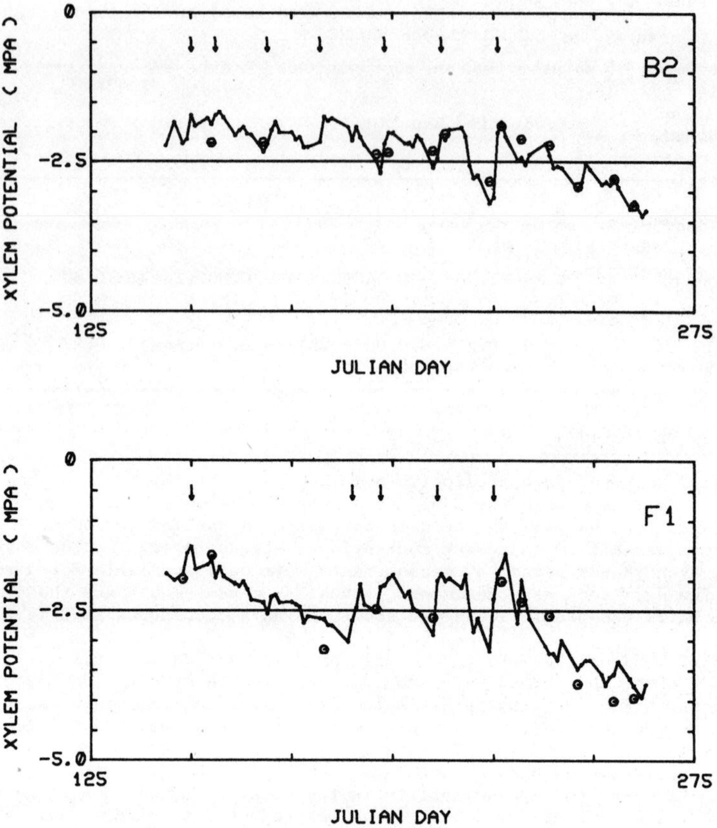

Fig. 3. Seasonal trends of predicted leaf xylem pressure potential (solid line) and actual measurements (circles) for two early season cotton irrigation treatments. Irrigation events are represented by arrows.

Acknowledgments

The authors gratefully acknowledge the technical assistance of Mr. Ron Seay and Mr. Pat Brooks.

REFERENCES

1. Ehrler, W. L. 1973. Cotton leaf temperatures as related to soil water depletion and meteorological factors. Agron. J. 65: 404-409.

2. Ehrler, W. L., S. B. Idso, R. D. Jackson, and R. J. Reginato. 1978. Wheat canopy temperature: Relation to plant water potential. Agron. J. 70: 251-256.

3. Geiser, K. M., D. C. Slack, E. R. Allred, and K. W. Stange. 1981. Irrigation scheduling using crop canopy - air temperature differences. Trans. Am. Soc. Agric. Eng. (in press).

4. Hiler, E. A. and R. N. Clark. 1971. Stress day index to characterize effects of water stress on crop yields. Trans. Am. Soc. Agric. Eng. 14: 757-761.

5. Hiler, E. A., T. A. Howell, R. B. Lewis, and R. P. Boos. 1974. Irrigation timing by the stress day index method. Trans. Amer. Soc. Agric. Eng. 17: 393-398.

6. Idso, S. B., R. D. Jackson, P. J. Pinter, Jr., R. J. Reginato, and J. L. Hatfield. 1981a. Normalizing the stress-degree-day parameter for environmental variability. Agricultural Meteorology 24: 45-55.

7. Idso, S. B., R. J. Reginato, D. C. Reicosky, and J. L. Hatfield. 1981b. Determining soil-induced plant water potential in alfalfa by means of infrared thermometry. Agron. J. (submitted).

8. Jackson, R. D., V. V. Salomonson, and T. J. Schmugge. 1980. Irrigation management - future techniques. Proceedings, Second National Irrigation Symposium, American Society of Agricultural Engineers, Lincoln, Nebraska, 20-23 October 1980, (in press).

9. Jackson, R. D. Canopy temperatures and crop water stress. In "Advances in irrigation," D. Hillel (ed), Academic Press, NY, (in press).

10. Kanemasu, E. T. (ed). 1975. Measurement of stomatal aperature and diffusive resistance. Wash. State Univ. Bull. 809, Pullman, WA. pp. 39.

11. Longenecker, D. E. and P. J. Lyerly. 1969. Moisture content of cotton leaves and petioles as related to environmental moisture stress. Agron. J. 61: 687-690.

12. Namken, L. N. 1965. Relative turgidity technique for scheduling cotton (Gossypium hirsutum) irrigation. Agron. J. 57: 38-41.

13. Misra, R. D. and P. C. Pant. 1981. Criteria for scheduling the irrigation of wheat. Expl. Agric. 17: 157-162.

14. Perrier, A. 1971. Leaf temperature measurement. p. 632-671. In "Plant photosynthetic production manual of methods." Z. Sestak, J. Catsky and P. Jarvis (eds.). Dr. W. Junk, N.V. Publishers, The Hague.

15. Pinter, P. J., Jr. Remote sensing of microclimatic stress. In "Biometeorolgy in Integrated Pest Management," J. L. Hatfield and I. J. Thomson (eds), Academic Press (in press).

16. Ritchie, G. A. and T. M. Hinckley. 1975. The pressure chamber as an instrument for ecological research. p. 165-254. In "Advances in ecological research." A. MacFadyen (ed.), Academic Press, London.

17. Scholander, P. F., H. T. Hammel, E. D. Bradstreet, and E. A. Hemmingsen. 1965. Sap pressure in vascular plants. Science 148: 339-346.

18. Stegman, E. D., L. H. Schiele, and A. Baur. 1976. Plant water stress criteria for irrigation scheduling. Trans. Am. Soc. Agric. Eng. 19: 850-855.

19. Wiegand, C. L., and L. N. Namken. 1966. Influences of plant moisture stress, solar radiation, and air temperature on cotton leaf temperature. Agron. J. 58: 582-586.

IRRIGATION SCHEDULING WITH THERMAL INFRARED AND SPECTRAL REMOTE SENSING INPUTS

J. L. Hatfield
Nonmember, ASAE

ABSTRACT

Field studies on grain sorghum (Sorghum bicolor L. Moench) have been conducted to evaluate the use of spectral reflectance and thermal infrared to the crop water use and crop water stress. These studies have been conducted for three years on a Yolo loam soil with various rooting volumes to limit the water availability. Data show that the stress-degree-day provides a valid indicator of crop stress, and the leaf-air temperature difference increases rapidly above zero when more than 65% of the available water is depleted. The leaf-air temperature difference is also related to leaf-water potential, with an increase above zero when the potential decreases below -12.5 bars. Vegetative indices, as developed from spectral reflectance, track the leaf area of the crop both in the growth stages and in changes caused by water stress but are a result of water stress and not a valid early warning method. There appears to be promise in the use of infrared thermometry to schedule irrigations, and it could easily replace other more labor and time-intensive methods, plus cover large areas quickly and reliably.

INTRODUCTION

Irrigation scheduling requires some method of assessing the water availability to the crop with sufficient lead time to provide for a water application. If the method used does not provide sufficient warning, the crop yields will be reduced because of lack of water, while the other extreme could result in too much water applied and a waste of water and energy. Methods currently used are soil moisture measurements, plant measurements, and evapotranspiration models. The role of remote sensing into these various approaches has been investigated by a number of groups (Jackson et al. 1977; Clawson and Blad 1981; Hatfield et al. 1980; Geyser et al. 1980).

Remote sensing of a plant canopy includes both reflected and emitted radiation. Kanemasu et al. (1977) showed how an evapotranspiration model could be developed with a leaf-area index (LAI) obtained from Landsat data. Jackson et al. (1977) proposed that a stress-degree-day, a comparison of midday canopy, T_c, and air temperatures, T_a, could be used to estimate water requirements in wheat. Ehrler et al. (1978) also found that leaf-water potential was related to a comparison of T_c - T_a. These data for wheat suggested that remote sensing inputs could be effectively utilized for evaluating crop stress and possibly irrigation scheduling.

Walker and Hatfield (1979) found that total water use during the reproductive stage of kidney beans could be related to the summation of stress-degree-days (SDD) in this growth period. Hatfield et al. (1980) suggested that 10-15 positive SDD's would be accumulated when 65% of the available water was extracted from the available rooting zone. In their data, they found that there was considerable daily variation due to meteorological changes. Idso

The author is: J. L. Hatfield, Biometeorologist, Department of Land, Air and Water Resources, University of California, Davis.

et al. (1981) suggested that the behavior of canopy and air temperature differences ($T_c - T_a$) were a linear function of the vapor pressure deficit (VPD) when the crop was transpiring at the potential rate. The form of the equation for grain sorghum is $T_c - T_a = 1.40 - 1.39$ VPD for data collected at Davis during 1981. When a crop was transpiring at less than potential, then the $T_c - T_a$ value would be displaced to the left of the baseline. Jackson et al. (1981) combined net radiation with the canopy, wet- and dry-bulb temperatures to develop an energy balance approach termed the crop-water stress index. This crop-water stress index closely paralleled the removal of water from root zone. These results suggest that remote sensing input into irrigation scheduling would be feasible and practical.

METHODS AND MATERIALS

Studies have been conducted on a Yolo loam soil at Davis, California, on grain sorghum (Sorghum bicolor L. Moench) to evaluate the relationship between water use, plant stress indicators, canopy temperatures and spectral measurements. These studies were conducted in 1978, 1979, and 1980 on Dekalb A28+ grain sorghum, planted in 75 cm rows at a population of 130,000 plants per hectare. Differential irrigations were imposed at different stages of growth. Beginning in 1979, a controlled rooting plot experimental site was developed which limited the rooting depth to 50, 100, and 150 cm, as compared to the natural profile of 300 cm. These plots were 5 m x 20 m in size. In 1979 and 1980, the differential irrigations were imposed on each rooting depth.

Measurements in each plot included soil moisture as measured by a neutron probe, canopy temperatures measured with an infrared thermometer, plant water status measurements with a pressure bomb or diffusion porometer, and spectral reflectance with a hand-held radiometer. Neutron soil moisture measurements were made twice weekly. Canopy temperatures were measured daily at 1330-1400 PST with a Barnes PRT-5[1/] infrared thermometer or a Telatemp AG-42[1/]. Spectral reflectance was measured with an Exotech model 100A[1/] with four wavebands, 0.5-0.6, 0.6-0.7, 0.7-0.8, and 0.8-1.1 µm. These measurements were made at solar noon on selected days throughout the growing season. Physiological measurements were made between 1300-1400 PST on selected days and in selected treatments.

Ancillary plant measurements of height, leaf area, dry weight and phenological stage were made throughout the growing season. Some measurements were also made of the net photosynthesis of the upper leaves of the canopy. Additional meteorological variables included dry- and wet-bulb temperatures at the time of the canopy temperature measurement.

RESULTS AND DISCUSSION

Water use in grain sorghum is negatively related to the accumulation of the stress-degree-days as shown in Fig. 1. As Walker and Hatfield (1979) had previously shown for kidney beans, this relationship represents a general confirmation that the stress-degree-day does indeed relate to water stress to which the crop is subjected. It is encouraging here that the 1978 and 1979 data fit into the same population. Data for 1980 from these experiments are not available due to prolonged periods of instrumentation failure and an incomplete record. With this type of relationship, it is reasonable to expect that the stress-degree-day approach would provide a method of irrigation scheduling; however, it is necessary that the time period be expressed in days rather than a complete phenological stage as shown in Fig. 1.

[1/] Mention of a specific tradename or produce does not imply endorsement or preferential treatment by the California Experiment Station.

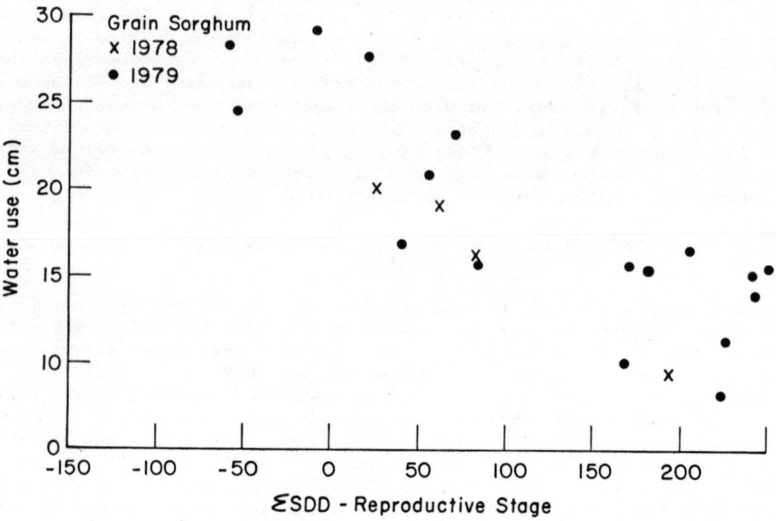

Fig. 1. Soil Moisture Use and the Summation of Stress-Degree-Days for the Reproduction Stage of Grain Sorghum in 1978 and 1979.

It is necessary to develop a relationship between the stress-degree-day and the water use of a crop, if an irrigation scheduling method is to be developed. In the Yolo loam soil with a natural rooting depth, a grain sorghum crop will grow roots deeper into the profile to extract water in order to avoid stress. The controlled rooting plots provided a methodology to control the rooting volume and thus allow for the evaluation of the stress-degree-day approach under a wider range of soil moisture conditions. Soil moisture and stress-degree-day data for two of the controlled rooting depth plots are given in Figs. 2 and 3.

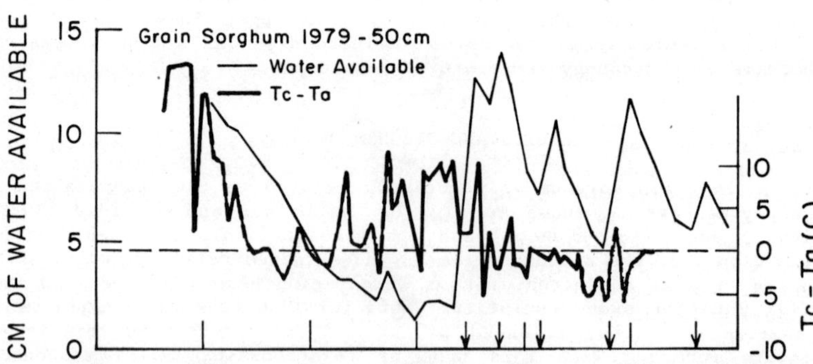

Fig. 2. Pattern of Seasonal Soil Moisture Use and T_c - Ta for Grain Sorghum Grown in a 50-cm Soil Profile in 1979.

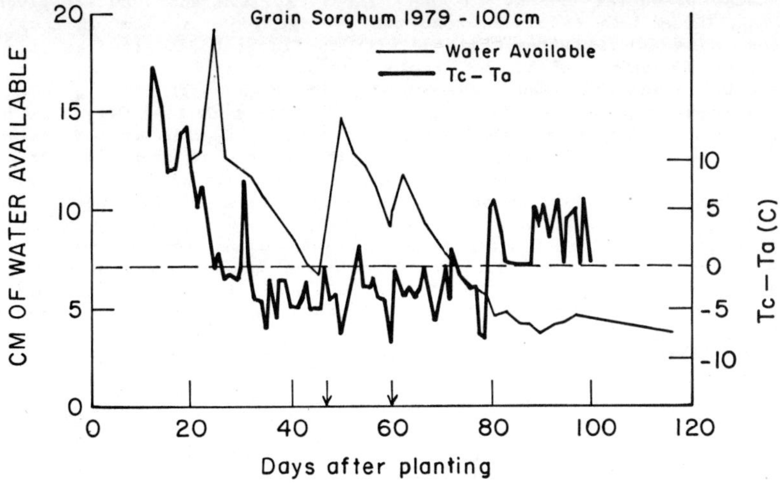

Fig. 3. Pattern of Seasonal Soil Moisture Use and T_c - T_a for Grain Sorghum Grown in a 100-cm Profile in 1979.

Data given in Fig. 2 represent the 50 cm plot that was maintained without water until 69 days after planting and then was maintained in a well-watered condition until maturity. Also shown on this graph is the daily values of the stress-degree-day throughout the growing season (Fig. 2). Early in the growing until 30 days after planting, the daily stress-degree-days are quite elevated due to the "warm" soil background relative to the leaf temperature. However, this portion of the growing cycle does not use much water if soil evaporation is kept to a minimum. It is evident that as the soil moisture became limiting, the daily stress-degree-day values were generally positive (Fig. 2). When the crop was rewatered at 79 days after planting, the stress-degree-day values decreased below zero and the canopy was cooler than air. The response of the crop was almost immediate after rewatering occurred. On Fig. 2, the zero line for T_c - T_a is placed at the extraction limit of 65% of the available water and in these data when the water availability dropped below this point, leaf temperatures remained elevated above air temperatures.

A similar effect is noted for a 100 cm rooting depth that was irrigated during mid-season and then allowed to dry out in the later stages of growth (Fig. 3). Again, as shown in Fig. 2, when 65% of the available water was extracted, the leaf temperatures became elevated above air temperatures. There was no indication of a leaf-temperature rise in the early portion of the season which could be due to a period of warm days with high vapor pressure deficits at this time. If the procedure suggested by Idso et al. (1981) compensates for environmental variability, then these values during this period would exhibit values of T_c - T_a below those that would normally be expected. Unfortunately a potential baseline for grain sorghum has not been developed and evaluated at this time.

All other plots for the 1979 and limited 1980 data exhibit these trends and, therefore, it is reasonable to expect that measurements of leaf temperature would be indicative of impending crop stress. From the data collected to date, a value of stress-degree-day that would be accumulated would become positive after 65% of the available water has been extracted. Once this level of depletion occurs, the daily values very quickly become positive. It is this aspect that needs refinement as suggested by Idso et al. (1981) and Jackson et al. (1981).

As a further evaluation of the stress-degree-day to quantify crop stress, the leaf-water potential and T - T data for 1978, 1979 and 1980 are given in Fig. 4. These data each represent the average of a minimum of four leaf samples. The fit is extremely good with the largest deviation at very low leaf-water potentials where the problems in attaining a good measurement are difficult. These data would suggest that in grain sorghum, as soon as the leaf becomes warmer than the air, the leaf-water potential decreases below -11.0 bars. Values below this potential have been used as indicative of water stress with prolonged exposure to this level of stress resulting in reduced growth and yield. These data suggest that days with leaf temperature above that of air would be an indicative of stress but does not provide a test of the crop stress approach suggested by Idso et al. (1981) because all these data were collected under a narrow range vapor pressure deficits from 2.0 to 4.0 kPa. Additional data collected on net photosynthesis and stomatal resistance for these same plots also exhibit a similar relationship as leaf-water potential. Net photosynthesis decreased rapidly as leaf temperatures increased above air temperature and stomatal resistance began to increase rapidly. These two parameters relative to T_c - T_a need additional research.

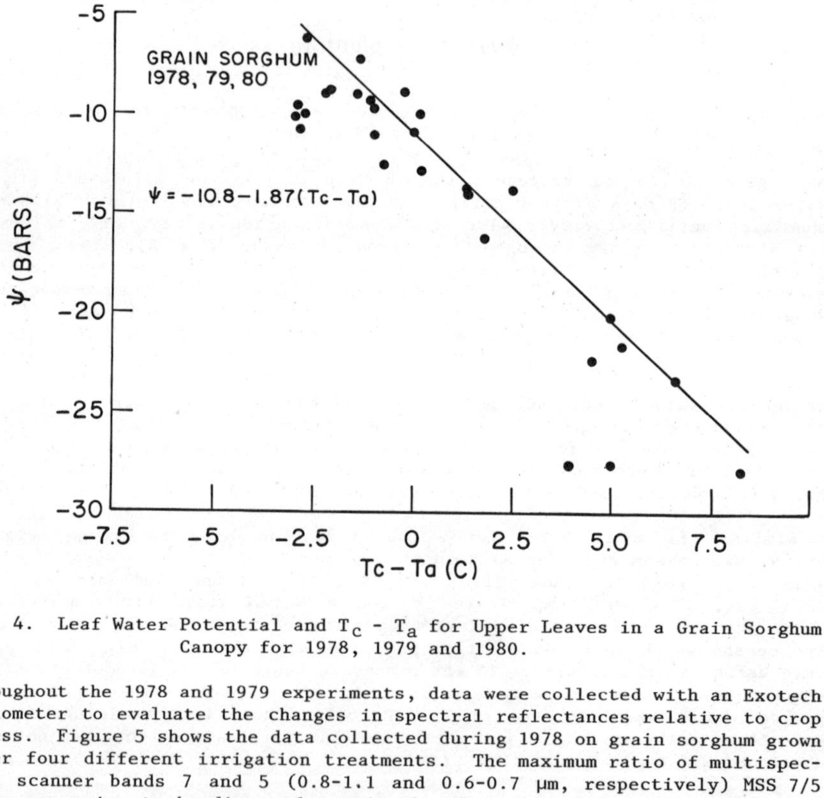

Fig. 4. Leaf Water Potential and T_c - T_a for Upper Leaves in a Grain Sorghum Canopy for 1978, 1979 and 1980.

Throughout the 1978 and 1979 experiments, data were collected with an Exotech radiometer to evaluate the changes in spectral reflectances relative to crop stress. Figure 5 shows the data collected during 1978 on grain sorghum grown under four different irrigation treatments. The maximum ratio of multispectral scanner bands 7 and 5 (0.8-1.1 and 0.6-0.7 μm, respectively) MSS 7/5 bands was prior to heading and was in the plot which had a substantial weed problem. Plots 3 and 4, which were not irrigated during the reproductive stage, showed a rapid increase in the rate of decline which is correlated with the rapid rate of leaf senescence (Fig. 5). Thus, the spectral reflectance patterns are very closely tied to the leaf area and the way in which the canopy intercepts radiation.

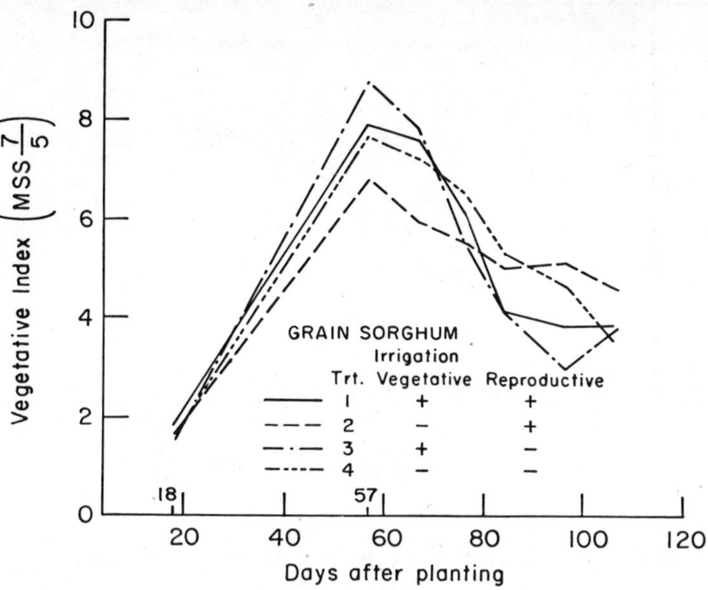

Fig. 5. Seasonal Patterns of Vegetative Index (MSS 7/5) for Grain Sorghum Under Four Irrigation Treatments in 1978.

To further the relationship between spectral refelectance and water stress with complete canopies a comparison was made of the vegetative index and the leaf-water potential of the upper leaves. Low leaf-water potential in grain sorghum are related to leaf rolling which should cause a change in the vegetative index. This change would result in less of the visible radiation being absorbed by the canopy and thus a decrease in the vegetative index. The only days considered in this analysis were those which had corresponding leaf-water potential data and spectral reflectance measurements; the results are given in Fig. 6. The data show that as the leaf-water potentials decrease below -15 bars, the vegetative index decreases. However, the change in the vegetative index is the result of water stress and not an early warning method. But as Kanemasu et al. (1981) have shown, it would be possible to use this method to evaluate the leaf-area index for evapotranspiration models.

CONCLUSIONS

The stress-degree-day concept is a valid indicator of crop stress as shown by the available soil-water and leaf-water potential data. Further refinements, such as those suggested by Idso et al. (1981) and Jackson et al. (1981), would appear to improve the capability of this technique into irrigation scheduling programs. The change in leaf-air temperature is closely related to the availability of water under the climatic conditions of California, which suggests a critical soil-water content when stress becomes detrimental to crop growth.

Spectral radiation as seen by reflected wavebands does follow the leaf area of the crop and under full cover the changes in crop canopies caused by leaf rolling are depicted in the vegetative index. This method does not provide a

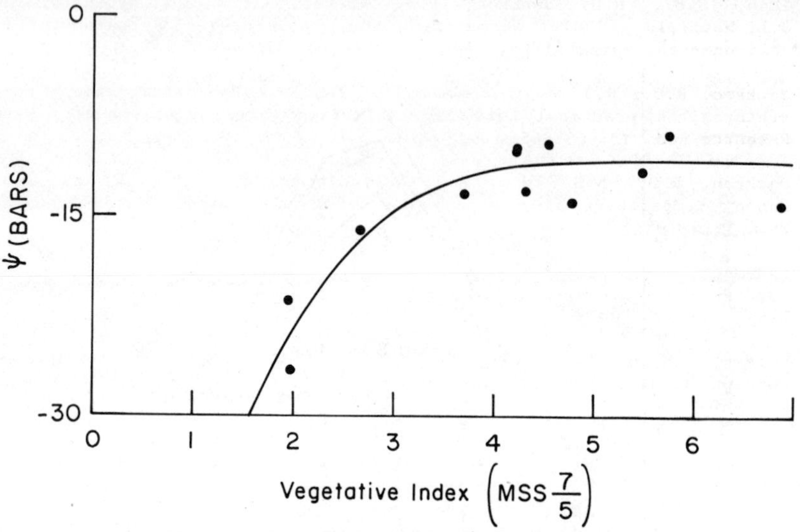

Fig. 6. Comparison of Leaf Water Potential and Vegetative Index of Grain Sorghum in 1979 with Fully Developed Canopy Subjected to Various Irrigation Treatments.

method of early warning but could be used as a method of assessing the impact on yield.

Recent studies utilizing the infrared thermometer in comparison with other methods at Davis show the stress-degree-day technique to be an effective means of scheduling irrigations. Further studies will continue to evaluate the technique for more crops and in comparison to other methods. Geyser et al. (1980) also found similar results but suggest that net radiation and vapor pressure deficit would be necessary data in more humid and variable areas. The refinement suggested by Idso et al. (1981) which incorporates the vapor pressure deficit needs evaluation in irrigation scheduling programs. Remote sensing inputs could be made easily and quickly over large areas allowing for a more complete examination of entire fields or farms.

LITERATURE CITED

1. Clawson, K.L. and B.L. Blad. 1981. Infrared thermometry for scheduling irrigations of corn. Agron. J. (in press).

2. Ehrler, W.L., S.B. Idso, R.D. Jackson and R.J. Reginato. 1978. Wheat canopy temperatures: relation to plant water potential. Agron. J. 70: 251-256.

3. Geyser, K.M., D.C. Slack, E.R. Allred and K.W. Stange. 1980. Irrigation scheduling using crop canopy-air temperature difference. ASAE Meetings, SW-599.

4. Hatfield, J.L., J.P. Millard, R.J. Reginato, R.D. Jackson, S.B. Idso, P.J. Pinter, Jr. and R.C. Goettleman. 1980. Spatial variability of surface temperature as related to cropping practice with implications for irrigation management. Proc. 14th ERIM Symposium on Remote Sensing of Environment. pp. 1311-1320.

5. Idso, S.B., R.D. Jackson, P.J. Pinter, Jr., R.J. Reginato, and J.L. Hatfield. 1981. Normalizing the stress-degree-day parameter for environmental variability. Agric. Meteorol. 24: 45-55.

6. Jackson, R.D., R.J. Reginato and S.B. Idso. 1977. Wheat canopy temperatures: a practical tool for evaluating water requirements. Water Resource Res. 13: 651-656.

7. Jackson, R.D., S.B. Idso, R.J. Reginato and P.J. Pinter, Jr. 1981. Canopy temperature as a crop water stress indicator. Water Resources Res. (in press).

8. Kanemasu, E.T., J.L. Heilman, J.O. Bagley and W.L. Powers. 1977. Using Landsat data to estimate evapotranspiration of winter wheat. Environ. Manag. 1: 515-520.

9. Walker, G.K. and J.L. Hatfield. 1979. Test of the stress-degree-day concept using multiple planting dates of red kidney beans. Agron. J. 71: 967-971.

CROP COEFFICIENTS FOR ESTIMATES OF DAILY

CROP EVAPOTRANSPIRATION

James L. Wright

Crop coefficients are used with values of reference evapotranspiration (ET) to estimate water use of a crop. Since the direct measurement of reference ET is expensive, time consuming, and laborious, it is usually preferable to calculate it from more easily obtainable climatic data. Extensive research on reference ET methods and improved crop coefficients has been conducted because of their application in irrigation scheduling and other aspects of water resources allocation, management and planning.

Various procedures have been used to obtain the necessary experimental crop and reference ET data, and several types of crop coefficient curves have been published during the past few years (Jensen 1974, Doorenbos and Pruitt 1977, Burman et al. 1980, Wright 1979, 1981). Crop coefficients must be matched with the appropriate reference ET. The climatic adequacy of the methods, the necessary input data, and the time scale all need to be understood and carefully applied if accurate estimates of crop water requirements are to be obtained for either irrigation scheduling or water resources planning. The available methods for estimating reference ET and improved crop curves, when properly applied, permit estimates of crop ET which are within the accuracy of most field irrigation systems to deliver water (Jensen et al. 1971, Jensen and Wright 1978, Wright and Jensen 1978).

This discussion briefly reviews the nature and origin of commonly used coefficients and outlines the conditions under which they can be appropriately applied. The application of recent "basal" crop coefficients (Wright 1981) is discussed and "mean" crop coefficients recently developed from ET data obtained with weighing lysimeters in Southern Idaho are also presented.

DEVELOPMENT OF CROP COEFFICIENTS

Crop Coefficients

Crop coefficients are generally empirical ratios of crop ET to some reference ET and are derived from experimental data. The time distribution of crop coefficients for a particular crop constitutes a "crop curve". A common form of the ET crop coefficient is:

$$K_c = E_{tc}/E_{tr} \qquad [1]$$

in which K_c = the dimensionless crop coefficient for a particular crop at a given growth stage and soil moisture condition, E_{tc} = the daily crop ET, and E_{tr} = the daily reference ET. Ideally E_{tr} characterizes the evaporative demand determined by meteorological conditions and K_c is a measure of the capability of the crop-soil-surface to meet that demand. Research has shown, however, that E_{tr} cannot be simply described for all climate and crop situations, partly because of the effects of the relative leaf area and the morphological and physiological characteristics of the crop canopy on the energy exchange and the aerodynamic diffusion processes within the atmosphere over a

The author is: Supervisory Soil Scientist, Snake River Conservation Research Center, USDA, Agricultural Research Service, Kimberly, Idaho 83341.

field. This complexity accounts for the many methods for estimating or predicting ET and the diversity of crop coefficients (Jensen 1974, Doorenbos and Pruitt 1977). Nonetheless, the goal of developing representative methods and coefficients for various ranges of climatic and crop conditions is warranted because crop coefficients provide a conservative means of estimating crop ET at progressive stages of growth.

The crop coefficient described by Eq. [1] includes effects of evaporation from both plant and soil surfaces, and is thus dependent upon soil water availability within the root zone and the wetness of the exposed soil surface. Soil evaporation is proportionally greater during the portions of the growing season when the crop canopy is at less than effective full cover. The time-scale of K_c is, of course, dependent upon that of E_{tc} and E_{tr}.

Changes in soil-water content with time are commonly used to obtain E_{tc}. Gravimetric sampling and neutron probe methods produce 3-5 day averages but even in carefully planned studies uncertainty exists concerning the significance of upward or downward movement of water and extraction by deep roots. Weighing lysimeters can provide daily ET data which are not subject to errors in assessing soil-water movement and the relative proportion of soil evaporation and transpiration can be estimated.

Reference ET

Various reference ET's can be used with Eq. [1] to develop crop coefficients. When these coefficients are subsequently used to calculate crop ET, the same type of E_{tr} should be used. In the past various forms of "potential ET", intended to describe near maximum ET, were used in developing crop coefficients. However, because of ambiguities in describing and interpreting potential ET, particularly in arid climates, the term "reference crop ET" with the reference crop specifically noted is recommended (Perrier 1979).

Doorenbos and Pruitt (1977) present detailed procedures for estimating E_{tr} for grass, which they define as "the rate of evapotranspiration from an extensive surface of 8-15 cm tall green grass cover of uniform height, actively growing, completely shading the ground, and not short of water". They list general crop coefficients adapted for use with grass E_{tr}.

Alfalfa E_{tr} was suggested for use in arid climates for irrigation scheduling procedures (Jensen et al. 1971). Alfalfa E_{tr} is defined as daily alfalfa ET when the crop occupies an extensive surface, is actively growing, at least 30-cm tall and standing erect, and is well watered so that soil water availability does not limit ET. Wright and Jensen (1972) used lysimeter ET data to develop procedures for estimating alfalfa E_{tr} from meteorological data. These procedures were later refined by Wright (1981). Alfalfa ET is usually greater than clipped grass ET, particularly under dry windy conditions, and may be preferable for arid regions because alfalfa is capable of near maximum ET rates when there is considerable advective sensible heat input from the air.

Estimating Crop ET

In practice, Eq. [1] is rewritten to solve for crop ET or E_{tr}:

$$E_{tc} = K_c \, E_{tr} \qquad [2]$$

where K_c is an experimentally derived crop coefficient and E_{tr} is estimated or measured in some appropriate manner. Problems sometimes occur in applying this concept because of misunderstandings or lack of attention to the details of the methods used in deriving K_c. The same type of E_{tr} should be used to estimate E_{tc} from climatic data as was used to determine K_c and the time-scale should be similar. When procedures and coefficients developed for a given climatic area are applied elsewhere, testing is advisable along with any adjustment of functional relationships as may be needed. A satisfactory degree of accuracy in estimating E_{tc}, i.e. ± 10%, generally requires careful adherence to the procedures and precautions to be discussed.

Methods available for estimating E_{tr} for use with Eq. [2] depend on data availability and local circumstances. Methods for estimating ET based solely on temperature are generally inadequate for arid or semiarid regions. Several suitable procedures for estimating irrigation water requirements have been reviewed in detail (Jensen 1974).

In developing guidelines for the determination of crop water requirements for application around the world, to fit various needs and availability of data, Doorenbos and Pruitt (1977) selected four methods for estimating E_{tr}, namely the (1) Blaney-Criddle, (2) radiation, (3) Penman combination equation and (4) pan evaporation methods. They developed correction coefficients for each method to adjust to a unified grass reference ET so that a single set of crop coefficients would suffice.

The Doorenbos and Pruitt version of the Blaney-Criddle method is considerably revised and for clarity could be referred to as the "FAO Blaney-Criddle" method to distinguish it from the original method of Blaney and Criddle (1950), and the widely used modified method developed by the USDA, Soil Conservation Service (1967). Versions of this method have been used extensively because of the relative availability of the data but the appropriate calibrations or adjustments have not always been applied. The monthly crop coefficients used with the original and SCS versions are different from the ET crop coefficient of Eq. [1] because a true reference ET is not used.

Doorenbos and Pruitt recommend the Penman combination equation method (Penman, 1948) for areas where measurements of temperature, humidity, wind and sunshine duration or radiation are available. While their procedures provide estimates of daily grass E_{tr}, other versions of the combination equation are available for estimating alfalfa E_{tr} (Wright and Jensen 1972, Wright 1981). Specific wind functions are recommended for local conditions for the most satisfactory results (Slatyer and McIlroy 1961). From a comparison of 16 methods for estimating ET at several locations around the world, the combination methods, with calibration of the wind and vapor deficit terms, agreed best with lysimeter-measured ET (see chap. VIII, Jensen 1974). The combination methods usually paralleled measured ET and displayed the best overall fit even without such calibration.

ET CROP COEFFICIENTS

Previous Crop Coefficients

In a sense, three sets of crop coefficients have been developed from the crop water requirement and irrigation scheduling research at Kimberly, Idaho, as data and methods were developed and needs changed. The crop coefficients originally used with the USDA-ARS Computerized Irrigation Scheduling Program (Jensen et al. 1971) were developed primarily from soil sampling data. These data, collected at various research locations, were generally for 5-15 day intervals. Each point on the crop curve, therefore, represented a mean for that period. Rapid changes such as occur with wetting or drying of the soil surface and crop development were damped. Contributions of evaporation from wet soil were partially included in the measured ET since the sampling period usually began 2 or 3 days after an irrigation or period of significant rainfall. The procedures for estimating E_{tr}, then referred to as potential ET and frequently noted as E_{tp} or E^*, were based on the combination equation with the wind function of Penman (1963).

Alfalfa Reference ET

Subsequent research indicated that the Penman wind function did not fit arid climatic conditions, so procedures were developed from measured net radiation and lysimeter ET data to permit estimating an alfalfa E_{tr} for arid regions (Wright and Jensen 1972). Lysimeter ET data were also utilized, as they became available, in revising crop coefficients which were provided to those using the scheduling program. A set of mean crop coefficients, based on alfalfa E_{tr} and denoted by K_{co}, were summarized for use in estimating irrigation water requirements for the type of crops grown in Southern Idaho (see Table 6.5, Jensen 1974). Some of these were essentially unchanged from the original coefficients of the scheduling program and some were revised according to results of lysimeter ET studies.

Since concurrent ET research had indicated that wet soil surfaces had a pronounced effect on ET, procedures were included in the USDA-ARS Irrigation Scheduling Program to adjust the crop coefficient for wet soil surfaces (Jensen et al. 1971). However, since the ET data used to develop the initial crop coefficients had also included some wet soil evaporation, this adjustment sometimes resulted in an overestimation of crop ET. Overestimation was also possible if the revised wind function of Wright and Jensen (1972) was used with the crop coefficients developed with the lower ET estimates of the Penman equation. Later comparisons with lysimeter ET data indicated overestimates early and late in the growing season with the developed procedures. Therefore, several years of lysimeter alfalfa ET and corresponding meteorological data were used to develop improved procedures for estimating alfalfa E_{tr} throughout the season (Wright 1981, 1982; see also Burman et al. 1980, 1981).

Basal ET Crop Coefficients

The revised procedures for estimating alfalfa E_{tr} and the lysimeter ET data available for most of the other irrigated crops typical of Southern Idaho, were used to develop a revised set of ET crop coefficients (Wright 1979, 1981). These were designed to represent dry soil surface conditions and were specifically called "basal ET crop coefficients", denoted by K_{cb} in following discussions, to emphasize this property and differentiate from previous versions. The experimentally derived values of K_{cb} were previously listed along with a similar set of basal ET crop coefficients developed by W. O. Pruitt from his lysimeter data and a grass E_{tr} for Davis, California (Burman et al. 1980).

Numeric adjustment of K_{cb} for the effects of surface soil wetness and soil drying properties can be accomplished with:

$$K_c = K_{cb} + (1 - K_{cb})[1-(t/t_d)^{1/2}](f_w) \qquad [3]$$

where K_c is the adjusted crop coefficient, K_{cb} is as previously defined, t = the number of days after major rain or irrigation, t_d = the usual number of days for the soil surface to dry, and f_w = the relative portion of the soil surface originally wetted. Values of K_{cb} for the times in question are used with this equation, which has a maximum value of 1. If irrigation is completed before noon, then t = 0 for that day; otherwise t = 0 for the following day. Data indicate that for heavy surface irrigations, f_w = 1.0 while for lighter irrigations, such as every other furrow and wetting only to the plant row, f_w = 0.5. For sprinkler irrigations or precipitation amounting to several days of ET, f_w = 1.0; otherwise progressively less. Kimberly ET data indicate that with bare soil or little crop cover at least 5 days elapse after major wetting of the soil surface before K_c returns to the basal level, thus t_d = 5 days or longer. For sandy soils, t_d is probably 3 or less and for clay loams probably 7 or more.

Since the decline in soil evaporation is related to the reciprocal of the square root of time, most of the adjustment by Eq. [3] takes place in the first few days after wetting the soil. Sprinkler irrigation completely wets the soil surface and results in higher ET following irrigation with partial crop cover than when row crops are furrow irrigated and the soil surface is only partially wetted. The general nature of the K_{cb} curve and examples of the adjustments provided with this equation are shown in Fig. 1. Some procedures for adjusting K_{cb} for the effects of limiting soil water have been presented previously (Jensen et al. 1971). Others could be developed for given soil and crop situations when data on the effects of inadequate soil water on ET are available.

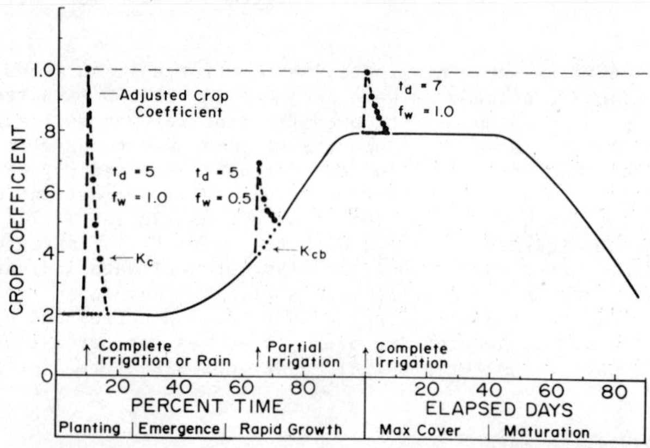

Fig. 1. Generalized basal ET crop coefficient (K_{cb}) curve showing means of calculating the crop coefficient (K_c) for wet surface soil by Eq. [3]

The use of the recent procedures for estimating E_{tr}, for either alfalfa or grass, and the new basal ET crop coefficients, along with procedures for accounting for extra wet evaporation and/or decreased transpiration due to limiting soil water, provides a consistent set of procedures for good estimates of crop ET where adequate data are available. Intermixing of coefficients and methods as they were developed along the way can lead to major errors.

NEW MEAN CROP COEFFICIENTS

Even though the basal ET crop coefficients allow for accurateestimates of daily crop ET, it is sometimes impractical to estimate wet soil effects and estimates of total seasonal water requirements are sometimes needed for a general area. Consequently, mean crop coefficients were developed using the ET data set obtained with lysimeters at Kimberly to meet these needs and are listed in Table 1. Corresponding crop development data are shown in Table 2. The mean crop coefficients of Table 1 are listed on a normalized time scale, similarly to the crop coefficients previously developed for the USDA-ARS Irrigation Scheduling Program (Table 6.5, Jensen 1974). Time from planting until full cover is considered on a percentage basis and time after full cover as elapsed days. This normalized time base helps fit a crop curve to different planting dates since large differences in planting date usually have only a minor effect on the date full cover is reached.

The data of Table 1 are for typical crop development and local management practices where root zone soil moisture does not limit crop growth and for

Table 1. Daily Mean ET Crop Coefficients (K_c), for normal irrigation and precipitation conditions, for use with alfalfa reference ET for crops grown in an arid region with a temperate intermountain climate. Coefficients were experimentally determined from weighing lysimeter ET data, Kimberly, Idaho, 1968-1978.

	Mean ET crop coefficients, K_c									
	Time from planting to effective cover (%)									
Crop	10	20	30	40	50	60	70	80	90	100
Barley	0.30	0.30	0.32	0.40	0.65	0.85	0.95	0.99	1.00	1.00
Peas	0.30	0.30	0.30	0.36	0.43	0.51	0.58	0.73	0.85	0.93
Sugar Beets	0.30	0.30	0.30	0.30	0.30	0.32	0.40	0.60	0.80	1.00
Potatoes	0.30	0.30	0.30	0.31	0.44	0.57	0.69	0.77	0.82	0.85
Corn	0.30	0.30	0.30	0.30	0.32	0.42	0.55	0.70	0.85	0.95
Beans	0.30	0.30	0.30	0.35	0.45	0.55	0.68	0.80	0.90	0.95
Winter Wheat	0.30	0.30	0.50	0.75	0.90	0.98	1.00	1.00	1.00	1.00
	Days after effective cover									
	10	20	30	40	50	60	70	80	90	100
Barley	1.00	1.00	0.90	0.50	0.25	0.15	-	-	-	-
Peas	0.90	0.65	0.53	0.35	0.20	0.15	-	-	-	-
Sugar Beets	1.00	1.00	1.00	0.98	0.94	0.89	0.85	0.80	0.74	0.60
Potatoes	0.85	0.83	0.81	0.79	0.75	0.70	0.65	0.50	0.35	0.25
Field Corn	0.96	0.95	0.94	0.90	0.85	0.79	0.74	0.35	0.25	-
Sweet Corn	0.93	0.93	0.90	0.85	0.75	0.58	0.40	0.20	-	-
Beans	0.95	0.90	0.67	0.33	0.15	0.10	-	-	-	-
Winter Wheat	1.00	1.00	1.00	0.95	0.55	0.25	0.15	0.15	-	-
	Time from new growth to harvest (%)									
	10	20	30	40	50	60	70	80	90	100
Alfalfa (1st)	0.70	0.82	0.91	0.96	1.00	1.00	0.98	0.96	0.95	0.95
(2nd & 3rd)	0.40	0.50	0.80	0.96	0.98	1.00	1.00	0.98	0.95	0.95
(4th)	0.40	0.44	0.60	0.65	0.55	0.50	0.45	0.35	0.30	0.25

Table 2. Dates of crop growth stages identifiable in the field for use with crop curves, Kimberly, Idaho, 1968-1978.

	Month/Day						Days	
	Planting	Emergence	Full Cover	Heading or Bloom	Ripening	Harvest	Planting to Full Cover	Full Cover to Harvest
Barley	4/4	4/15	6/20	6/20	7/15	8/10	80	55
Peas	4/10	4/25	6/05	6/15	7/05	7/25	55	50
Sugar Beets	4/15	5/10	7/10	-	-	10/15	85	95
Potatoes	4/25	5/25	7/10	7/01	9/20	10/10	75	90
Field Corn	5/5	5/25	7/15	7/30	9/10	9/20	72	70
Sweet Corn	5/5	5/25	7/15	7/20	8/15	8/15	72	30
Beans	5/22	6/05	7/15	7/05	8/15	8/30	55	45
Winter Wheat*	(2/15)	(3/01)	6/5	6/5	7/15	8/10	(110)	60
Alfalfa (1st)#	4/01					6/15		76
(2nd)	6/15					8/01		46
(3rd)	8/01					9/15		46
(4th)	9/15					10/30		46

*Effective dates in parenthesis. Crop planted on 10/10 and emerged 10/25.
#Effective planting date for established alfalfa is date growth begins in spring or harvest of preceding crop. Final harvest is date crop becomes dormant.

the rainfall and irrigation patterns of the area. The usual precipitation pattern consists of light rains during the spring with a dry summer averaging about 20, 25, 25, 5, 13, 13 and 16 mm per month for the April-October period, respectively. Five or more irrigations are required throughout the season for most crops. The mean K_c's of Table 1 can be used directly in Eq. [2] without adjustment for soil wetness by Eq. [3]. For the early portion of the season before major crop development, a mean coefficient of 0.3 represents the average conditions experienced at Kimberly including soil evaporation from residual winter season soil moisture, spring precipitation, and preplant or pre-emergence irrigations. When wetter than normal soil conditions exist throughout the period, a general coefficient of 0.4 is appropriate until the crop emerges and begins rapid growth. For unusually dry conditions, a value of 0.2 is indicated.

The several alfalfa harvests are considered individually because of the climatic differences for each growth period. The growth period is from new growth, or harvest to harvest. Thus K_c for alfalfa reaches 1.0 at a normalized time scale value of 50%. Alfalfa K_c declines slightly before harvest because of lodging of the crop. The mean K_c values of Table 1 were developed from the same data base as the previously published K_{cb} values (Wright 1979, 1981) and are, of course, generally higher during the period of crop development and at full cover, because of the inclusion of wet soil evaporation.

In the case of winter wheat, the procedure used in normalizing the time scale for the other crops is not appropriate. Therefore, an effective planting date of mid-February and an effective emergence of about March 1 was assumed as shown in Table 2. This corresponds quite closely to the time when winter wheat passes from the cold weather dormant condition to greenup and the beginning of rapid leaf area expansion.

Dates for crop development stages easily recognizable in the field are listed for the various crops in Table 2, since it is often difficult to determine when a crop reaches effective full cover. In the case of winter and spring cereals, the most easily detectable stage is that of heading. Normal dates of heading are usually known in an area. If there are seasonal differences, adjustments can be made. For the row crops: beans, sugarbeets, potatoes and corn, the effective full cover date is when the leaves of plants in adjacent rows intermingle so that soil shading becomes nearly complete. If this does not occur, the maximum value of K_c should be scaled down. During this period of crop development, leaf area expansion is very rapid and the exact day of effective cover is not critical. Winter wheat, spring cereals, and sugarbeets at full cover had crop coefficients similar to full cover alfalfa. The other crops peaked at values of about 0.95, except for potatoes which peaked at about 0.85. Most of the crops showed a gradual decline after the peak because of lodging or leaf senescence. With maturation there was a rapid decline in K_c. The low potato value is assumed to be due to physiological differences of the plant as it was an excellent crop. However, others have found higher values for potatoes (Doorenbos and Pruitt 1977, Burman et al. 1980).

PAN EVAPORATION

Because of the interest in using pan evaporation measurements to estimate crop ET, the evaporation from the U. S. Class A pan maintained by the U. S. Weather Service at Kimberly was compared with alfalfa E_{tr}, as shown in Fig. 2. This pan is sited in an irrigated grass plot about 40 m square and is surrounded by irrigated fields planted to various crops each year. The daily ratios plotted in Fig. 2 are 7-day means of daily values for a 14-year period and thus represent considerable averaging. The ratio gradually decreased in early spring as the grass around the pan site began to grow and ET increased in surrounding fields. It was fairly steady at about 1.1 during the central portion of the growing season and gradually increased in late summer

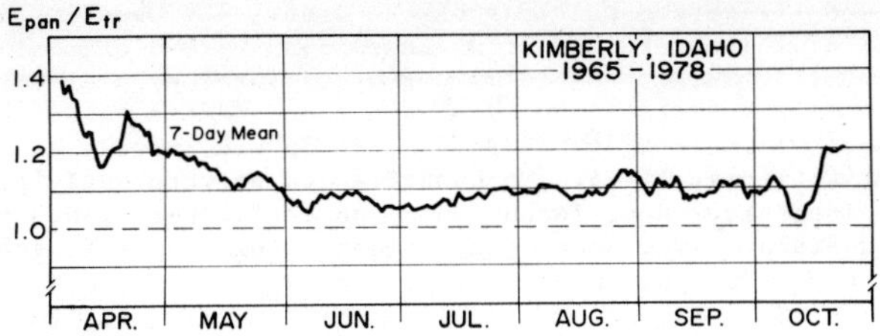

Fig. 2. Seven-day running means of the ratio of class A pan evaporation (E_{pan}) to alfalfa E_{tr} for a 14-year period.

and early fall as surrounding crops matured and conditions favored increased sensible heat advection from surrounding dry areas. The reciprocal of this ratio provides a "pan factor" to adjust pan evaporation to alfalfa E_{tr}. However, since a pan and a crop respond differently to daily changes in meteorological conditions, this method provides averages probably only suitable for periods of 10 days or longer. The recommendations and procedures for using evaporation pans as summarized by Doorenbos and Pruitt (1977) should be closely followed if pans are to be used.

APPLICATION OF ET CROP COEFFICIENTS

The improved ET crop coefficients developed from lysimeter data should be usable in estimating crop ET in areas with climates similar to that of south-central Idaho because the crop coefficient is a relative factor and differences in water use due to climate can be accounted for in the computed reference ET. They should also be usable in areas with different climates if verified procedures are used to estimate the reference ET or if correction factors are used to adjust this reference. While there is some variation in the rate of crop development at various locations and on different years, because of seasonal differences, the crop curve can be shifted to account for this variation if a few simple crop development characteristics are occasionally monitored. The newer crop curves developed from lysimeter data should be used with the appropriate recently developed procedures for calculating reference crop ET. When properly applied, the newer crop coefficients are expected to increase the accuracy of irrigation scheduling procedures and the estimates of crop water use from historical records.

REFERENCES

1. Blaney, H. F., and W. D. Criddle. 1950. Determining water requirements in irrigated areas from climatological and irrigation data. U. S. Dept. of Agric., Soil Conservation Service Technical Bulletin 96, 48 pp.

2. Burman, R. D., P. R. Nixon, J. L. Wright, and W. O. Pruitt. 1980. Water requirements. Pages 189-232 in M. E. Jensen, ed., Design and Operation of Farm Irrigation Systems. Amer. Soc. Agric. Engr., St. Joseph, Michigan.

3. Burman, R. D., J. L. Wright, P. R. Nixon, and R. W. Hill. 1981. Irrigation management - Water requirements and water balance. Proc. Amer. Soc. Agric. Engr., Second National Irrigation Symposium, Oct. 1980: 141-153.

4. Doorenbos, J., and W. O. Pruitt. 1977. Guidelines for predicting crop water requirements, FAO Irrig. and Drain. Paper 24 (rev.) 156 pp.

5. Jensen, M. E. (ed.). 1974. Consumptive use of water and irrigation water requirements. Rep. Techn. Comm. on Irrig. Water Requirements, Irrig. and Drain. Div., Amer. Soc. of Civil Engr., 345 East 47th Street, New York, N.Y. 227 pp.

6. Jensen, M. E., and J. L. Wright. 1978. The role of evapotranspiration models in irrigation scheduling. TRANSACTIONS of the ASAE 21(1):82-87.

7. Jensen, M. E., J. L. Wright, and B. J. Pratt. 1971. Estimating soil moisture depletion from climate, crop and soil data. TRANSACTIONS of the ASAE 14(6):954-959.

8. Penman, H. L. 1948. Natural evaporation from open water, bare soil, and grass. Proc. Roy. Soc. London. A 193:120-145.

9. Penman, J. L. 1963. Vegetation and Hydrology, Techn. Communication No. 53, Commonwealth Bureau of Soils, Harpenden, England, 125 pp.

10. Perrier, A., Chairman. (M. E. Jensen, English version). 1979. Question 1 - Terminology, and Question 2 - Measurement Methods. Pages A-120 - A-145 <u>in</u> Ann. Rpt. ICID International Round Table Conference on Evapotranspiration, Budapest, Hungary, 26-28 May, 1977. Intern. Comm. on Irrig. and Drain., 48 Nyaya Marg, Chanakyapuri, New Delhi, India 110021.

11. Slatyer, R. O., and I. C. McIlroy. 1961. Practical Climatology. CSIRO, Melbourne, UNESCO, Paris, 340 pp.

12. U. S. Department of Agriculture, Soil Conservation Service. 1967. Irrigation Water Requirements. Technical Release No. 21, 83 pp.

13. Wright, J. L. 1979. Recent developments in determining crop coefficient values. Proc. Irrig. and Drain. Div. Specialty Conference, July 17-20, 1979, p. 161-162. (Abstr., also mimeo. tech. rpt., 13 pp.)

14. Wright, J. L. 1981. New evapotranspiration crop coefficients. J. Irrig. and Drain. Div., Amer. Soc. of Civil Engr. (In press)

15. Wright, J. L. 1982. Alfalfa reference evapotranspiration calculated from meteorological data in arid irrigated areas. (In review)

16. Wright, J. L., and M. E. Jensen. 1972. Peak water requirements of crops in southern Idaho. Proc. Amer. Soc. Civil Engr. of the Irrig. and Drain. Div. 98(IR2): 193-201.

17. Wright, J. L., and M. E. Jensen. 1978. Development and evaluation of evapotranspiration models for irrigation scheduling. TRANSACTIONS of the ASAE 21(1):88-96.

DESKTOP DATA SYSTEM FOR IRRIGATION SCHEDULING

Craig E. Crouch 1/, William E. Hart 2/,
Gary D. Jardine 3/, Ronald J. Brase 4/

The authors have written a program for a desktop data processing system which allows irrigations to be scheduled by an operator with no technical knowledge of computer operation or irrigation scheduling. The program utilizes the expanded capabilities of desktop microcomputers and the techniques of interactive, user-oriented programming to allow its use by a person with minimal clerical skills, under the supervision of an agronomist or agricultural engineer. Much of the flexibility and ease of use of the program results from its division into six "modules", or program segments, each performing a major processing task. Each module is brought into the computer's memory from a magnetic tape as it is needed.

PROGRAM STRUCTURE - THE SIX MODULES

The six program modules are stored on one of the two cassette tapes which comprise the microcomputer's offline permanent storage (the second tape is reserved for data.) The SUPERVISORY module assumes overall program control (Figure 1) when the machine is switched on. Each time the user selects a major processing task, the SUPERVISORY module copies the appropriate program module into memory.

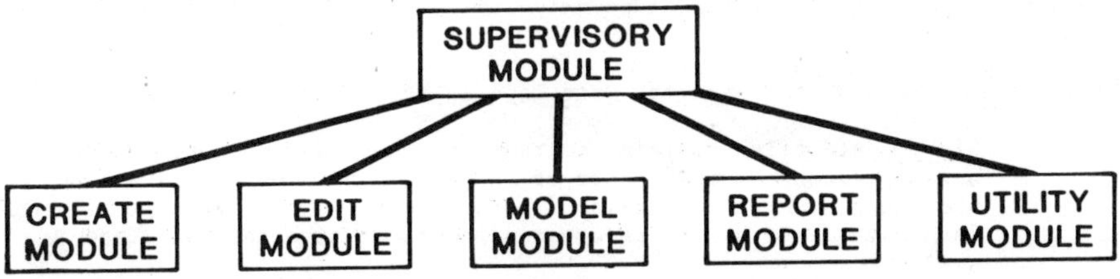

Fig. 1. Structure of Program Control

The specific functions of each module are described below in the typical order of their use:

SUPERVISORY Module--Brings other modules into memory and transfers data to and from tape.

CREATE Module--Adds new data files to the Data Tape which are used in the scheduling operation.

1/ Water Resources Engineer, CH2M HILL--Engineers, Planners, Economists, and Scientists, Sacramento, CA.
2/ Principal, Water, Waste, and Land, Inc., Ft. Collins, CO.
3/ Agricultural Engineer, CH2M HILL, Redding, CA.
4/ Principal, Crop Care Associates, Fresno, CA.

EDIT Module--Allows the user to make changes to the data files, including addition and revision of weather data and the dates and amounts of measured depletions and actual irrigations.

MODEL Module--Performs evapotranspiration (ET) calculations for each field using the information from the data files and calculates the projected irrigation schedule date. Results are stored on the Data Tape.

REPORT Module--Generates any of four types of reports using the data files updated by MODEL.

UTILITY Module--Carries out a variety of supporting functions, such as creating backup data files and plotting crop curves.

USER FEATURES AND PROGRAM OPERATION

The Hewlett-Packard System 45 microcomputer was selected for the desktop data system because of its operating convenience and flexibility. This machine has a cathode ray tube (CRT) for data display, a thermal printer, two cassette tape drives for storing data and programs, and 64K bytes of usable memory. Data entry is accomplished using the typewriter keyboard and numeric key pad. The keyboard also has 16 special function keys (SFKs) for which special operations may be predefined; these keys are used to select tasks to be performed by the program. The Extended BASIC programming language of the System 45 is capable of linking the individual program modules, a feature essential in providing a program which actively assists the operator.

Magnetic Tapes and Data Files

The machine's two magnetic tape drives allow the operator to quickly access and store a very large amount of information. The Program Tape is inserted in the right-hand tape drive and holds all of the program modules. The Data Tape is placed in the left-hand tape drive and contains all of the data files for the program.

Data and results are stored on the Data Tape in three types of files:

- Global data files contain program data, such as the most recent access sequence number, date of the last backup of the data tape, and other supervisory items. There is one such Global data file on each Data Tape, and it is created the first time the Data Tape is used.

- Regional data files hold data specific to a group of farms. Regions are selected by the supervising agronomist so that all farms within a region may be assumed to have weather characteristics which are identical in all respects except rainfall. Most of the regional data is needed to calculate daily potential evapotranspiration. A region file name (identifier) is made up of one to six characters.

- Farm data files hold data specific to a farm (the region to which it is assigned, rainfall, etc.), and all data (such as soil, crop, and irrigation information) and scheduling results for up to 20 individual fields on that farm. A farm file name (identifier) has the same structure as a region file name.

Farms may be switched from one region to another by editing (changing) the assigned region name. Each Data Tape is wholly independent, so there is no maximum number of Regions or Farms that can be scheduled by the user.

Program Operation

When the microcomputer is switched on, the supervisory portion of the program is automatically transferred from tape into the computer memory and takes control. Thereafter, the program prompts the operator to enter responses or data with clear, step-by-step instructions and queries. Tasks may be selected by the user by use of special function keys (SFKs), or by "menu" selections.

Special function keys are labeled with abbreviations describing their various tasks. The program will display a list of SFKs and will request the user to press one to select a program module or a processing task. Only special function keys which apply within a portion of the program are effective, protecting the operator from incorrect selections. The last SFK in each list, labeled END, ends processing using that list.

Very often a menu (Figure 2) of numbered tasks will be displayed on the CRT and the item number of the desired task requested. Item numbers entered which do not correspond to a task on the menu are ignored by the program and another item number is requested. If no menu item is entered, processing using that menu will end and a new SFK task selection will be requested.

```
TRIAL Field Data Menu :
   1 = Crop
   2 = Soil
   3 = Emergence date
   4 = Full cover date
   5 = Irrigation cutoff date
   6 = Depletion
   7 = Irrigation
   8 = Allowable depletion
   9 = Root zone depth
  10 = ETP Multiplier
  CONTinue will End field editting and Schedule the field
```

Fig. 2. A Typical Processing Menu

The use of the SFK lists and menus results in a division of tasks by processing levels. If a problem is encountered, this structure allows the user to "back-out" of processing by selecting the END SFK or not entering a menu item. In addition, when data or a yes/no response is requested and none is entered, the program will return to a more general processing level, allowing the user to retain control even if he does not know a correct response.

Other features of the program check selections and entries by the operator to detect errors and allow for their correction. For example, actions by the operator which would destroy data require confirmation. Also, numerical data as they are entered are checked against a range of allowable values established by the supervising agronomist. The data system will notify the operator when a range is exceeded, display the entered value, and then allow the user to revise the value or confirm that it is correct. When climatic data are entered, the user will be warned if dates are out of sequence, the date needed next will be displayed, and the user will be asked to re-enter the date. Throughout, the program will respond to incorrect entries by prompting the user to supply the appropriate response.

Minimal knowledge and effort is necessary for an operator to access and store data. Data files are selected by entering a region or farm identifier up to six characters long; fields are selected from within farm data files by a similar identifier. When the operator chooses not to select a region, farm, or field, the corresponding data file last accessed will be used, considerably speeding use of the system.

When files are being substantially changed, it is often advantageous to save results or changes at intermediate points with a single keystroke using the SAVE SFK. But in any case, the system automatically saves changes to data files on tape before proceeding to the next task.

IRRIGATION SCHEDULING OPERATIONS

Prescheduling Operations

The value of potential evapotranspiration (ETP) is computed for each day of new weather data by three different methods--the Jensen-Haise (Jensen 1974), the Penman combination, and the Pruitt-modified Penman (Doorenbos and Pruitt 1977)--with progressively more detailed weather data required for each method. The associated ETP is recomputed when weather data are revised and the user may specify that this new ETP value be reflected in the next scheduling run. The historical mean ETP curve (used for estimating evapotranspiration beyond the latest weather date) can be edited, as can the Jensen-Haise coefficients for the region.5/ Rainfall may be entered for each farm and changed if an error is noted. Effective rainfall is computed from the gross amount entered.

Field data may be revised as the growing season proceeds. Actual or allowable depletions may be entered either as percents or amounts. Changes to a field's root zone depth or to the six 1-foot incremental water holding capacities will automatically be reflected in a revision of the allowable depletion. Fields of a farm may be listed and reordered for the reports. They may be removed from scheduling, or scheduled but selectively printed so they do not appear on client reports.

Scheduling of Irrigations

Irrigation scheduling is performed for all fields of each farm selected.6/ Regional daily ETP computed using any of the three equations may be used for scheduling. The historical mean ETP curve is used to calculate the depletion of soil moisture beyond the latest day for which weather is available. This forecasted ETP may be adjusted, using a multiplier, to reflect weather trends. Starting with a complete irrigation or a measured soil depletion and accounting for irrigations and rainfall, the program calculates daily soil depletions for each field until the allowable limit is reached. The date and amount of the needed irrigation is logged on the printer and stored back on tape for inclusion in reports. Crop growth curves for up to 40 different crops and crop varieties may be defined in the program.

Fourteen days of weather data are stored and each field should be scheduled at intervals of 10 days or less, but scheduling may be resumed at any time by entering a current measured depletion or a complete irrigation for a field. Fields may be passed over for scheduling (made inactive) by entering a negative crop number. The results of all scheduling calculations are stored on the Data Tape.

5/ The coefficient curves (Kc for crop age, Ks for soil water stress, and Ke reflecting excess evaporation after a rain or an irrigation) necessary for estimating crop evapotranspiration on a given day are defined by the program but can also be edited as necessary.

6/ Intermediate values of all curve data (e.g., mean historical ETP, Ks, Ke) are calculated using a central-difference interpolation polynomial.

Trial Schedules

A special TRIAL mode allows the operator to investigate various irrigation schedules for individual fields. In this mode, limited editing capabilities enable the user to vary most field parameters and assess different irrigation strategies. A summary TRIAL report is displayed on the CRT, or may be printed out. After concluding all trials, the user has the option of permanently saving the results. This feature essentially provides a manual procedure for optimizing the irrigation schedule of each field.

Reports

All reports are generated from the data stored on the Data Tape. There are four different report formats available.

The Client Soil Moisture Report (Figure 3) includes a summary of the climatic data, the calculated ETPs used for scheduling, and irrigations and depletions for each field. The most important information for the client, the next scheduled irrigation date, is highlighted for quick reference. Also printed are averages of computed ETP for the past 14 and 7 days, and mean daily averages for the next 7, 14, and 28 days.

The Climatic Summary details all of the weather data for the past 14 days, along with averages for each weather data item. ETPs for this period are displayed for all three equations for comparison, and the ETP averages are again provided.

The Detailed Soil Moisture Report presents all the parameters for each field on a given farm. A marked-up copy of this report is normally employed by the user to update the data for each field. Information typically updated includes such items as extensions in root zones, new irrigation events, and new measured depletions. Calculated ET data presented include crop coefficient (Kc), the soil water stress coefficient (Ks), and the surface moisture evaporative losses coefficient (Ke). In addition, evapotranspiration (ET) is projected for the next 7 days and to the next irrigation date for each field.

The Field Information Report (Figure 4) presents all of the field parameters required for irrigation scheduling, including water holding capacities, root zone depth, and summaries of the season's irrigations.

A large number of reports can be printed without an operator standing by. The names of regions, farms, and even individual fields for which reports are needed are "stacked up" in queues, along with the report types. After the desired reports are specified, an SFK is selected which will cause printing of either all, or only "active," fields. The operator may use the HALT SFK to stop printing and add paper. The machine will issue an accelerating beep to signal the end of report printing.

ADDITIONAL FEATURES

Record Keeping

Upon entry to the program, the operator is required to supply the date. This date and a session number appear at the top of the first page printed in that program run, and are reprinted when each module is called. Complete records (logs) of file creations and edits are automatically printed as these tasks are performed. A Run Log is also created, summarizing scheduling results for each field as it is processed by the MODEL module. At any point, the user may interrupt processing by pressing the COMMENT SFK and type comments which are directly printed on the Logs. Printing of these Entry, Change, and Run Logs can be avoided at the user's option by selecting the CRT SFK (which merely shifts output to the CRT from the printer). However, coupled with the

C R O P   C A R E   A S S O C I A T E S

GOLDEN WEST FARMS

*** 6/12/81 ***

Climatic Information:

Date	Temperatures		Relative Humidity		Solar	Wind Run	Rain	Potential Evapotranspiration
Mo/Day	Max	Min	Max	Min	Radiation	Miles	In	(In/Day)
5/27	91	60	100%	37%	648	76	0.00	0.303
5/28	95	59	100%	29%	673	73	0.00	0.323
5/29	97	58	89%	23%	670	87	0.00	0.349
5/30	94	62	60%	22%	670	72	0.00	0.370
5/31	94	55	96%	11%	679	54	0.00	0.326
6/1	97	62	79%	31%	614	38	0.00	0.303
6/2	91	63	64%	39%	675	130	0.00	0.389
6/3	96	59	67%	20%	688	73	0.00	0.371
6/4	101	61	80%	22%	702	106	0.00	0.397
6/5	106	66	73%	21%	702	69	0.00	0.392
6/6	99	67	60%	22%	672	92	0.00	0.405
6/7	96	59	45%	13%	707	103	0.00	0.437
6/8	97	64	79%	29%	697	100	0.00	0.382
6/9	92	59	76%	24%	702	110	0.00	0.383
Avergs:	96	61	76%	25%	679	85		0.366

Evapotranspiration Potential:

Average ETP Has Been:
- For Last 7 Days 0.395 In/Day
- For Last 14 Days 0.366 In/Day

Predicted ETP Is:
- For Next 7 Days 0.314 In/Day
- For Next 14 Days 0.319 In/Day
- For Next 28 Days 0.316 In/Day

Individual Field Information:

Field		Recomended Next Irrig		Root Zone	Soil Depletions		Seasonal Irrigations		Last Irrigation	
Name	Crop	Date	Amt	Feet	Allowable	Current	No.	Total	Date	Amt
3-1	Tomato	6/15	4.3	3.0	50%	23%	6	15.7	6/5	3.1
3-2	Tomato	6/13	4.4	3.0	50%	33%	6	18.5	6/4	2.4
4-3Est	Tomato	6/15	4.2	3.0	45%	46%	5	13.5	6/9	2.1
2-1	Cotton	6/19	4.2	2.5	60%	43%	1	8.1	12/7	8.1
3-3	Cotton	6/19	4.3	2.5	60%	43%	1	9.0	12/4	9.1
3-4	Cotton	6/19	4.2	2.5	60%	43%	1	9.0	12/4	9.1
4-1	Cotton	6/16	4.4	2.5	60%	49%	1	9.5	12/16	9.5
4-2	Cotton	6/18	4.5	2.5	60%	45%	1	9.5	12/10	9.5
4-4	Cotton	6/17	4.4	2.5	60%	47%	1	9.5	12/14	9.5
34-2	Cotton	6/16	4.6	2.5	60%	50%	1	7.4	11/29	8.0
35-1	Cotton	6/22	2.5	2.0	45%	35%	1	5.0	12/1	6.7
35-2	Cotton	6/24	2.4	2.0	45%	33%	1	8.4	11/30	8.4
35-3	Cotton	6/22	2.4	2.0	45%	35%	1	3.7	12/3	3.7
35-4	Cotton	6/24	2.4	2.0	45%	34%	1	9.1	12/3	9.1

Fig. 3. Client Soil Moisture Report

Region: MURSTA	GOLDEN WEST FARMS	Date: 06.12.81
Forecast: 100% Normal	(CLKCAN)	Sess No: 47 PG:1

Identification:						
Field Name	3-1	3-2	4-3Est	2-1	3-3	3-4
Field Code	1	2	3	4	5	6
Acreage	155	155	80	155	155	155
Crop Elements:						
Crop Name	Tomato	Tomato	Tomato	Cotton	Cotton	Cotton
Curve No	16	16	16	1	1	1
Emergence	4.05.81	4.05.81	4.01.81	5.09.81	5.06.81	5.07.81
Full Cover	6.10.81	6.10.81	6.01.81	7.25.81	7.25.81	7.25.81
Irr Cut-Off	7.15.81	7.15.81	7.15.81	8.30.81	8.30.81	8.30.81
Soil Elements:						
Root Zone-Ft	3.0	3.0	3.0	2.5	2.5	2.5
WHC - In	6.3	6.3	6.6	5.3	5.3	5.3
Soil Type	2	2	2	2	2	2
Drainage Days	1	1	1	1	1	1
Allowable Depl:						
Percent	50.0	50.0	45.0	60.0	60.0	60.0
Inches	3.15	3.15	2.97	3.15	3.15	3.15
Depletions:						
Verified	6.08.81	5.31.81	6.08.81	5.17.81	5.17.81	5.17.81
Current:						
Date	6.07.81	6.07.81	6.07.81	6.07.81	6.07.81	6.07.81
Percent	22.9	32.7	45.5	43.4	43.2	42.5
Inches	1.44	2.06	3.00	2.28	2.27	2.23
Past:						
Date	6.07.81	6.07.81	5.31.81	5.17.81	5.17.81	5.17.81
Percent	22.9	32.7	47.0	18.0	15.0	15.0
Inches	1.44	2.06	3.10	.95	.79	.79
Future:						
Date	0	0	0	0	0	0
Percent	0.0	0.0	0.0	0.0	0.0	0.0
Inches	0.00	0.00	0.00	0.00	0.00	0.00
ET Elements:						
Date	6.07.81	6.07.81	6.07.81	6.07.81	6.07.81	6.07.81
K-Crop	0.945	0.945	1.001	0.334	0.360	0.351
K-Stress	0.979	0.954	0.688	0.681	0.685	0.693
K-Excess	0.000	0.000	0.000	0.000	0.000	0.000
ET-Current	0.405	0.394	0.301	0.100	0.108	0.107
Average ET:						
Next 7 Days	0.230	0.000	0.220	0.072	0.077	0.076
To Next Irr	0.224	0.205	0.218	0.073	0.077	0.077
Irrigations:						
Efficiency-%	.75	.75	.75	.75	.75	.75
Last:						
Date	6.05.81	6.04.81	6.02.81	12.07.80	12.04.80	12.04.80
Gross-In	3.1	2.4	2.2	8.1	9.1	9.1
Net - In	2.3	1.8	1.7	6.1	6.8	6.8
Future:						
Date	0	0	6.09.81	0	0	0
Gross-In	0.0	0.0	2.1	0.0	0.0	0.0
Net - In	0.0	0.0	1.6	0.0	0.0	0.0
Next Scheduled:						
Date	6.15.81	6.13.81	6.15.81	6.19.81	6.19.81	6.19.81
Refill(Net)	3.2	3.3	3.1	3.2	3.2	3.2

Fig. 4. Field Information Report

backup features (described below) the logs could prove invaluable, both in the case of Data Tape loss or damage, and by leaving a clear work trail for review by the supervising professional. The logs are especially useful when performing TRIAL scheduling calculations, providing a record of the effects of the irrigation schedules examined.

Backup Files

Because there is always some possibility of tape damage or accidental loss of entered data, it is advisable to keep backup Data Tapes. The data system will print a message at the end of the processing session when a duplicate data tape should be made. Since 14 days of weather data are needed for scheduling, a backup interval of 7 days was selected. The 50-percent overlap of data on consecutive backup tapes reduces updating effort by at least half if data must be re-entered. In addition, the program will print a backup warning message after a specified number of sessions. The session backup interval value set in the program is 25, but may be altered to suit the needs of the user. Finally, a directory of all data files may be generated using the DIRECTORY SFK, and the UTILITY module may be used to easily create copies of all files on the Data Tape.

Documentation

Documentation for the data system includes a 50-page User's Manual which clearly explains and illustrates all operational steps, and a 140-page Programmer's Manual which contains all necessary supporting technical data.

PROGRAM APPLICATION

In the past 2 years, the desktop data system has been successfully applied by Crop Care Associates in scheduling irrigations for a variety of crops in the San Joaquin Valley of California. In general, the program has proven to be trouble-free, easy to use and understand, and a valuable aid in crop water management. An average of 1 hour of preparation and 3 hours at the computer were required each week to schedule irrigations for 75 individual fields on five farms.

While the program was developed for an irrigation scheduling consultant, it can be applied with equal success by corporate farming interests, grower cooperatives, and individual growers, and by irrigation/water districts.

The needs of corporate farming interests, grower cooperatives, and in some cases, individual growers can justify purchase of a sophisticated irrigation scheduling system. Qualified in-house personnel can directly apply the program to meet each organization's scheduling needs as an alternative to hiring an agronomic consultant to provide this service. In each case, the organization's particular needs would determine whether the investment in the computer equipment, program, and training of staff can be justified and offset by savings of water, labor, and energy and by increased agricultural production.

Irrigation/water districts can make the desktop data system available for direct use by their growers, where the growers have a need but cannot individually afford the investment. Because the desktop system is easy to operate, each grower is responsible for data collection, scheduling, and field verification, with the irrigation district staff or local extension personnel providing occasional technical support. Most important, each grower receives the benefits of irrigation schedules which utilize personal knowledge of his farming operation. Districts can achieve water and energy conservation while improving crop production by providing an irrigation scheduling capability, without assuming the major labor costs associated with complete scheduling services.

SUMMARY

Using this desktop data system, irrigation scheduling may be accomplished by a person with minimal clerical skills. The detailed processing logs and the variety of reports provided aid in review and supervision of this work by an agronomist or agricultural engineer. No prior experience with computers is needed by the user.

Application of the scheduling system to irrigated agriculture can be tailored to meet the needs of the individual user or group of users. Simplicity of operation combined with built-in sophisticated irrigation scheduling techniques allow the desktop data processing system to be successfully applied to situations where previous irrigation scheduling approaches have failed.

REFERENCES

1. Doorenbos, J. and W. O. Pruitt. 1977. Crop water requirements. FAO Irrigation and Drainage Paper No. 24. Food and Agriculture Organization of the United Nations. Rome. 144 p.

2. Jensen, M. E. (ed.) 1974. Consumptive Use of Water and Irrigation Water Requirements. American Society of Civil Engineers, New York, N.Y. 216 p.

3. Jensen, M. E., J. L. Wright and B. J. Pratt. 1971. Estimating Soil Moisture Depletion from Climate, Crop and Soil Data. TRANS. of the ASAE 14: 954-959.

4. Kincaid, D. C. and D. F. Heermann. 1974. Scheduling Irrigations Using a Programmable Calculator (ARS-NC-12). Agricultural Research Service, U.S. Department of Agriculture. 55 pp.

IRRIGATION MANAGEMENT STRATEGIES FOR HUMID REGIONS

L. C. Hammond R. B. Campbell E. D. Threadgill
Member ASAE

The need to irrigate many humid region soils for high crop yields has been well documented (Bruce et al. 1980, Butson and Prine 1968, Sheridan et al. 1979, van Bavel and Verlinden 1956, and van Bavel and Carreker 1957). Growth in irrigation farming in the southeastern United States has outpaced the accumulation of research data necessary for the development of irrigation practices which maximize profits. However, a growing worldwide demand for food and fiber coupled with an energy crisis has stimulated research on efficient irrigation management. In humid regions, where irrigation water is not necessary for salt control, the goal is to maximize the evapotranspirational use of irrigation and rainfall while minimizing the leaching loss of water, fertilizers, and pesticides during the crop-growing season. A near ideal water situation would be a soil water profile near maximum capacity at planting, but depleted by 50% or more at harvest, and a season-long rainfall and irrigation distribution pattern producing no plant water stress and no drainage loss. Crop yields have been shown to increase linearly with actual evapotranspiration (ET) until potential ET has been attained (deWit 1958, Hanks 1974, Musick and Dusek 1980, Skogerboe et al. 1979, and Tanner 1981). This response is not surprising in view of the fact that the same stomatal barrier is encountered by both CO_2 and water vapor during photosynthesis and transpiration. Low stomatal resistances are necessary for photosynthetic processes to function at optimum levels within constraints of other factors (solar radiation, temperature, nature and growth stage of crop, and CO_2 levels).

A series of strategies will be necessary to approach the above goal of 100% efficiency in irrigation water use. The simple question of when to irrigate and how much water to apply leads to a myriad of factors and decision-making consequences. We use the term "strategy" to encompass the complex and dynamic decision-making process which leads to a series of irrigation scheduling practices appropriate to the ever-changing environmental and economic conditions encountered by the farmer in producing a crop.

The purpose of this paper is to discuss strategies in irrigation scheduling resulting from recent experiences in the southeastern United States.

METHODS

An irrigation management field experiment on corn was established in 1979 at Gainesville, Florida, on Lake fine sand--an excessively drained, coated, thermic, Typic Quartzipsamment. Funk G-4507 corn hybrid was planted on March 13, 1979 in 90-cm rows at a population of 71,000 plants/ha. A broadcast application of 4-3.5-13.3 (NPK) fertilizer (1120 kg/ha) was incorporated in

---

The authors are: L. C. HAMMOND, Professor, Soil Science Department, University of Florida, Gainesville, FL; R. B. CAMPBELL, Soil Scientist, Coastal Plains SWCRC, Florence, SC; and E. D. THREADGILL, Associate Professor, Department of Agricultural Engineering, Coastal Plain Experiment Station, University of Georgia, Tifton, GA.

the top 10 cm of soil one week ahead of planting. Two sidedressings with NH_4NO_3 gave a total N application of 200 kg/ha. A solid set, impact sprinkler system delivering 2.54 cm of water per hour was used to apply the desired irrigation quantities during early morning hours when winds were calm. The plots were 13.7 m x 13.7 m in size and arranged in a randomized complete block design with four replications. Water management treatments were: (1) rainfed; (2) irrigation, light rate (0.8-1.9 cm); (3) irrigation, medium rate (1.7-2.5 cm); (4) irrigation, light rate, and only after beginning tassel; and (5) irrigation, light rate, and only up to beginning tassel. The maximum irrigation rates were used after day 68. Irrigation was scheduled on treatment 2 by a water budget procedure[1/] using calculated (Jensen and Haise 1963) daily ET rates. Irrigations were scheduled on treatments 3, 4, and 5 when pore-water pressures became more negative than minus 200 mb at 15-cm depths. In all irrigation treatments, the basic strategy was to refill part of the depleted soil profile, thus leaving some storage to be filled by rainfall. However, no effort was made to maintain a constant or even a known allowable depletion except in the model scheduling of treatment 2 where the allowable depletion was varied from 95% the first two weeks to 50% by midseason and 60% the last two weeks (based on the developing root zone).

Response to water management was determined as yield of corn grain. A simple water balance simulation (Rao et al. 1976, 1981) provided estimates of daily and seasonal ET and drainage, and daily soil profile water contents. Required model inputs include calculated daily ET rates, measured soil water characteristics (water contents at field capacity and permanent wilting, hydraulic conductivity and water redistribution time), and root depth with time. In the water balance simulations, daily ET values were obtained from monthly potential ET rates calculated by the Penman method from long-term weather records and handbook tables of extraterrestrial radiation. An arbitrary 10% downward adjustment of the ET rate was made for an incomplete crop canopy (0-25 days) during the early part of the season, and a 10% upward adjustment was made for later in the season (after 40 days). Measured field capacity and permanent wilting volume percentages were 7.5 and 2.2, respectively, to the 28-cm depth, and 6.5 and 2.2 from 28 cm to maximum root depth at 180 cm. The maximum soil water content during infiltration was set at 20% by volume and a 5-day redistribution time was used.

YIELDS AND WATER BALANCE

Rainfall and irrigation distributions for 3 of the 5 treatments are shown in Fig. 1. In addition, the estimated daily water losses by drainage from the 180-cm soil profile are shown. Changes in the comparative water inputs and the drainage losses reflect increasing rooting depth with time. Irrigation treatments 2 and 3 produced drainage at times when no drainage occurred from the rainfall treatment. Grain yields and complete seasonal water balance data for all treatments are given in Table 1. Drainage, ET, and profile water depletion are estimates from the simulation procedure. Grain yields were low and variable due to fertility problems associated with earlier experiments on the field site. Although the relationship of yield and seasonal irrigation and ET is less precise than in some of our other studies, the overall findings will be useful to our discussion of irrigation management strategies.

The effect of ET, irrigation, and rainfall over a 7-day period on pore water pressure in the profiles of treatments 2 and 3 is shown in Table 2. These data show the effect of treatment on the initial water pressure distribution, depth of restoration of the depleted profiles by different irrigation amounts, and depths of wetting following rainfall. Treatments 2 and 3 had

---

1/ See Lambert et al. elsewhere in these Proceedings.

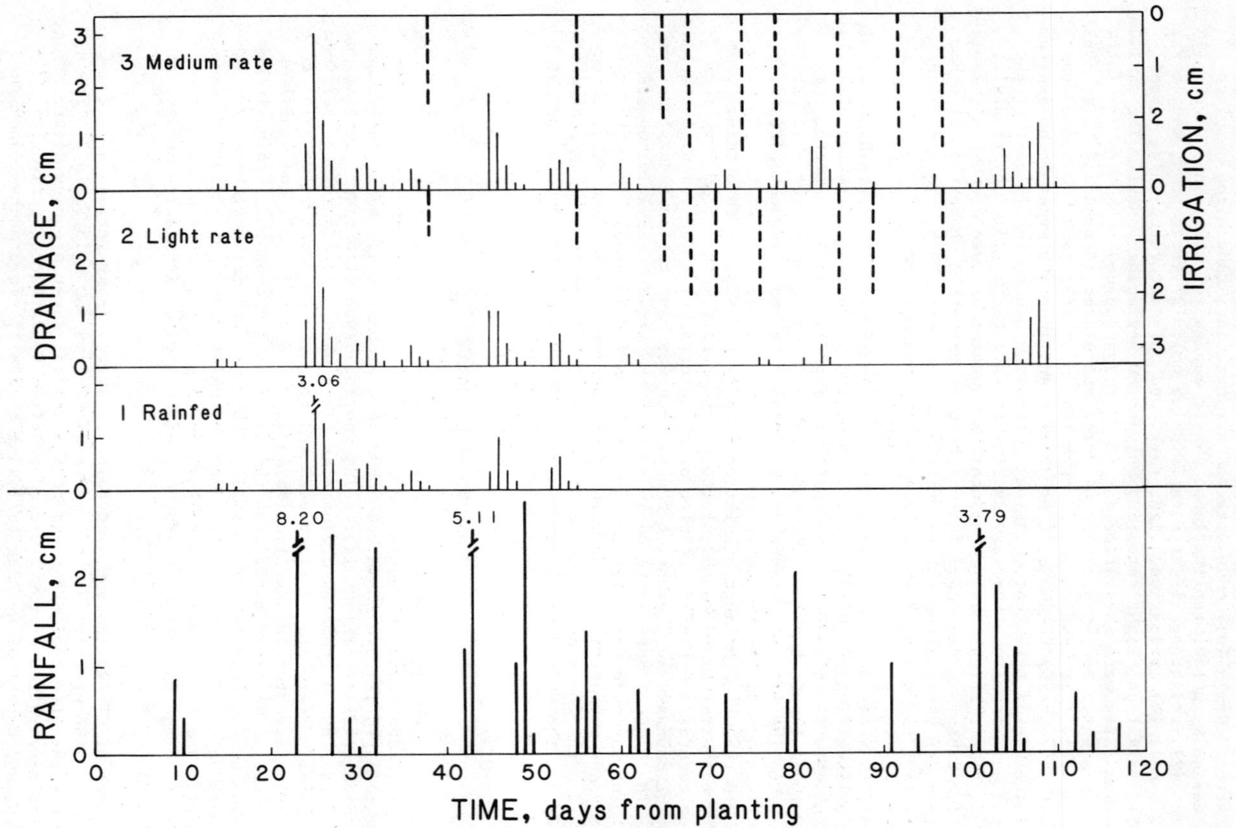

Fig. 1. Rainfall, Irrigation, and Drainage (from 180 cm Profile) in Three Water Management Treatments on Corn, Gainesville, 1979.

been irrigated 8 and 5 days earlier, respectively. The calculated available water depletion on June 17th was 44% and 27% for treatments 2 and 3, respectively, assuming a root depth of 164 cm. Differences in depth of wetting on June 24 and the resultant differences in drainage from subsequent rainfalls as shown in Fig. 1 (days 100 to 110) illustrate the water-conserving value of a partial refilling of the depleted profile with each irrigation.

Table 1. Effect of Water Management on Seasonal Water Balance Components and Yield of Corn, Gainesville, 1979

Treatment	ET	Irrig.[a]	Soil water depletion[b]	Drainage	Grain yield
	(cm)				(kg/ha)
(1) Rainfed	36.8	-	6.1	12.7	1430
(2) Irrigated, light	42.9	14.6	3.1	18.2	5120
(3) Irrigated, medium	42.9	20.6	3.1	24.2	5190
(4) Irrigated, stress[c]	42.9	12.8	3.1	16.4	4250
(5) Irrigated, stress[c]	37.5	2.1	5.9	14.1	2690

[a]Rainfall, 43.4 cm.

[b]Soil profile water content to maximum root depth at planting (12 cm) minus content at crop maturity.

[c]Irrigation started only after beginning tassel on treatment 4 and terminated at that time on treatment 5.

Table 2. Soil Water Pressure Distribution under Two Irrigation Treatments on Three Dates in 1979

	Soil water pressure (negative)[a]					
	Treatment 2			Treatment 3		
Depth	6/17	6/19	6/24	6/17	6/19	6/24
(cm)	(millibars)					
15	354	75	77	445	67	98
30	170	65	68	114	51	76
45	128	147	61	79	70	73
60	126	135	60	67	70	66
75	131	151	58	68	77	59
90	152	192	216	356	502	58
120	-	-	126	-	-	44
150	-	-	103	-	-	43

[a]Irrigation of 1.9 and 2.5 cm on treatments 2 and 3, respectively, on June 18. Rainfall: 3.8 cm on June 21 and 1.9 cm on June 23.

Grain yield response to irrigation and to ET is shown in Fig. 2. The regression coefficients are similar to those found by us in other experiments, as well as those reported by others (Musick and Dusek 1980, Morey et al. 1980, Rhoads and Stanley 1973, Robertson et al. 1973, Skogerboe et al. 1979, and Stewart et al. 1975). The relationships shown in Fig. 2 are apparently typical of crop response to water supply (Stegman et al. 1980) and will be used as a focal point for our discussion of irrigation strategies.

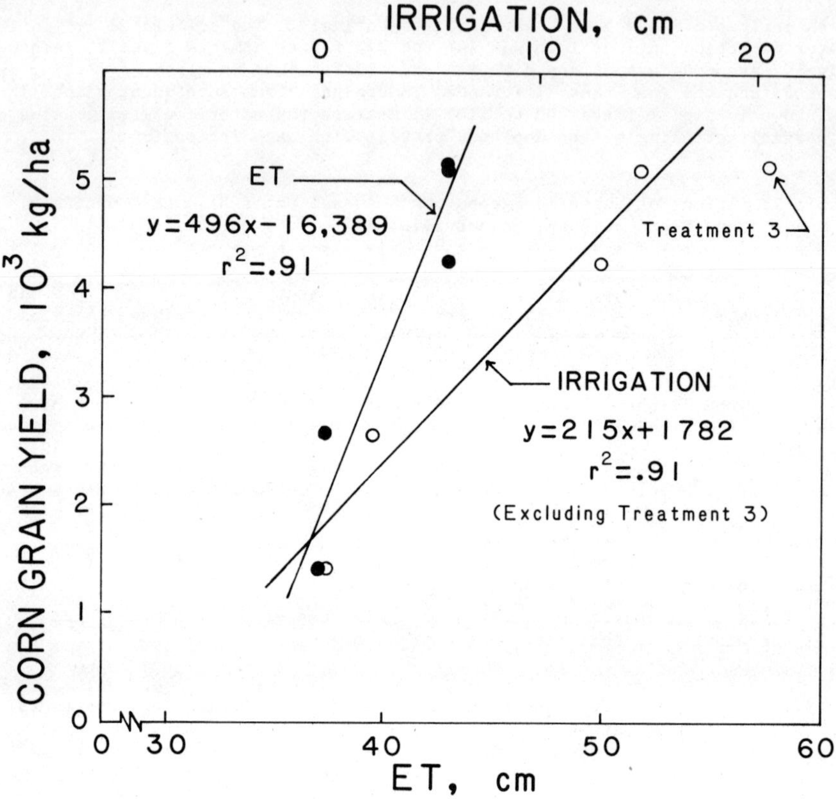

Fig. 2. Grain Yield in Relation to Seasonal Irrigation and Simulated ET, Gainesville, 1979.

The relationships in Fig. 2 are in agreement with theoretical expectations. If all applied water was used as evapotranspiration, the two curves would be superimposed. It does not appear likely that the irrigation function would ever be larger than the ET function. The smaller slope of the irrigation function is an indication of the inefficiency of irrigation management for the particular season. In fact, an estimate of the fraction of applied water which was used as ET is given by the ratio of regression coefficients (21.5/496 = 0.43). Likewise, individual treatment estimates are given by the ratio of the increase in ET and the seasonal irrigation amount (for treatment 3, 6.1/20.6 = 0.30). The latter calculations are more appropriate if, as indicated by Stegman et al. (1980), the irrigation function is typically curvilinear and curves away from the ET function as irrigation depths increase. In 1980 Georgia experiments, an estimated 67% of irrigation was used to increase ET. A comparable value of 70% was obtained in Florida in 1980 from irrigation and ET production functions of 350 and 500 kg/ha-cm, respectively. Because of rainfall uncertainty in the southeastern U.S., it is not possible to avoid an appreciable loss of irrigation water by deep percolation.

Production functions like those in Fig. 2 are useful in several ways. First, the general relationship of crop yield to irrigation must be determined through experience and research in order to develop economical water management systems and practices for particular soil-climate-crop systems. The researcher investigates the processes affecting the relationship and tests experimental systems and practices against the historical data base. The farmer develops management strategies to achieve what has been shown to be possible and more. Policymakers and planners need the above information in developing and allocating resources. For humid conditions, it appears that the linear portion of the irrigation function should be the working basis for irrigation scheduling strategies. The goal should be to achieve the yield potential associated with maximum ET while minimizing non-ET losses of applied water. As others have shown (Phene and Beale 1976, Rawlins and Raats 1975, Skogerboe et al 1979, and Stegman et al. 1980) and as the data in Table 2 and Fig. 2 show, relative to treatment 3, irrigation must be scheduled frequently and at rates which leave some unfilled soil storage for following rainfall. Non-ET losses of water may interact with the crop and soil to cause yield reductions through leaching of nutrients and poor soil aeration (Campbell and Phene 1977, Campbell and Moreau 1979, and Skogerboe et al. 1979).

The ET function can be a family of curves depending on the sequencing of ET deficits during the growing season (Stegman et al. 1980). The latter also affects the scatter of data points for the irrigation function. However, a mean ET function based on near optimum ET deficit sequencing is a valuable function. It represents a goal and a standard against which strategy results can be tested. Moreover, this function as well as others in the family of functions predicts the yield reduction for given ET deficits.

Water management implications of these two types of functions have been discussed in detail by Stegman et al. (1980). It is encouraging that recent results of water management studies conducted in the humid southeastern U.S. parallel the more extensive data from subhumid and semiarid regions and can be subjected to similar analysis and evaluation. Differences in water management goals and strategies between these regions should, in the main, reflect differences in available water resources and rainfall uncertainties.

Additional water management strategy considerations will be mentioned only briefly. Strategies related to fractional replenishment of the depleted soil profile and deficit irrigation cannot be isolated from the question of irrigation timing. A strategy of variable timing throughout the crop season would be expected to be most appropriate to the temporal changes in canopy coverage, plant rooting depth, and plant growth stage-related sensitivity to water stress. Timing may be based on the plant condition (water potential, observable wilting, etc.), soil condition (water potential, water content, and water depletion level and distribution in the root zone), and estimated ET (from pan evaporation or from real time of historical meteorological data, solar radiation, vapor pressure, wind speed, and temperature). Crop condition is the basic consideration since the goal is to supply the water needs of the plant for maximum photosynthesis. In the absence of measurements of crop and soil conditions, observation of indicator plants (weeds, sensitive varieties or cultivars) or indicator areas in the field can be helpful in determining when to irrigate.

Adapting the cropping pattern and sequence to irrigation farming is another aspect of management strategy. High capital and operating costs of irrigation equipment may require changes such as multiple cropping--a practice currently being recommended in Georgia as a means of reducing the risk of irrigation farming.

System design impacts vary strongly on irrigation strategies directed toward the goal of full ET for the plant canopy and no drainage of water below the root zone. A high degree of application uniformity is a necessity. Thus, irrigation may need to be scheduled at night when wind is less of a factor. Although many systems are large (100 ha or more), smaller center-pivot systems (20 ha or less) are becoming more common throughout the southeastern

U.S. due to small field size and rolling topography. These smaller systems may require 24 hours or less per revolution while large systems may require 48 hours or more. Both are usually designed to replace an expected accumulated ET depth during a revolution. Frequently, the speed (and thus, output) can be varied to meet variable ET demands and permit short shut-down periods between revolutions. A system not able to meet ET demand during the designed revolution time forces a deficit irrigation practice which will not supply water needs during an extended drought. For the March-June period in South Georgia, Sheridan et al. (1979) have found that in one out of 2 years there will be at least 21 consecutive days with less than 0.6 cm of rainfall on any day. Attempts to minimize underdesigned system problems by beginning irrigation ahead of need results in water loss during rainfall intervals and water stress damage during long droughts. Special soil and climate conditions in Nebraska permitted the successful use of high-frequency irrigation at less than ET demand rates (Fischbach and Somerhalder 1974). However, discussions with Cooperative Extension Service personnel and farmers revealed that substantial yield losses were experienced in 1977, 1980, and 1981 in Georgia due to delaying the initial irrigation for 2-3 days and then applying water at ET demand rates during an extended drought. The soil profile initially became so dry that the system simply could not "catch up." Consequently, only the upper portion of the root zone had adequate moisture for the duration of the drought. This is typical of center-pivot irrigation systems (Stegman et al. 1980), and more research data are needed to resolve this particular design and management problem under humid conditions.

CONCLUSIONS

There are two production functions for consideration in developing irrigation strategies and in evaluating the successes or failures of strategies used. The ET function (yield per unit of water used in seasonal ET) should be reasonably constant from season to season and represent a maximum or goal to achieve in irrigation management. Moreover, this function predicts the losses in yield to be expected per unit of seasonal ET deficit. In marked contrast, the irrigation function (yield per unit of irrigation water applied during the season) is highly variable with season and with the irrigation treatments used to find the function. Nevertheless, average irrigation function values (in the linear range) are needed in assessing the economic potential of irrigation farming. Essential elements of irrigation management strategies for humid regions include: variable scheduling practices with stage of growth and crop condition, irrigation to replenish a fraction of the water-depleted profile, irrigation to prevent plant water stress, and the development of an intensive overall crop management scheme attuned to the dynamic soil-plant-climate system.

REFERENCES

1. Bruce, R.R., J.L. Chesness, T.C. Keisling, J.E. Pallas, Jr., D.A. Smittle, J.R. Stansel and A.W. Thomas. 1980. Irrigation of crops in the southeastern United States: Principles and practice. USDA-SEA. ARM-S-9/May 1980. 54 p.

2. Butson, K.D. and G.M. Prine. 1968. Weekly rainfall frequencies in Florida. Florida Ag. Exp. Sta. Circular S-187. April 1968. 41 p.

3. Campbell, R.B. and R.A. Moreau. 1979. Ethylene in a compacted field soil and its effect on growth, tuber quality, and yield of potatoes. Amer. Potato J. 56: 199-210.

4. Campbell, R.B. and C.J. Phene. 1977. Tillage, matric potential, oxygen and millet yield relations in a layered soil. TRANS. of the ASAE 20: 271-275.

5. deWit, C.T. 1958. Transpiration and crop yields. Versl. Landbouwkd. Onderz. 64.4. 88 p.

6. Fischbach, P.E. and B.R. Somerhalder. 1974. Irrigation design requirements for corn. TRANS. of the ASAE 17: 162-165, 171.

7. Hanks, R.J. 1974. Model for predicting plant yield as influenced by water use. Agron. J. 66: 660-664.

8. Jensen, M.E. and H.R. Haise. 1963. Estimating evapotranspiration from solar radiation. J. Irrig. Drain. Div., ASCE 89: 15-41.

9. Morey, R.V., J.R. Gilley, F.G. Bergsrud and L.R. Dirkzwager. 1980. Yield response of corn related to soil moisture. TRANS. of the ASAE 23: 1165-1170.

10. Musick, J.T. and D.A. Dusek. 1980. Irrigated corn yield response to water. TRANS. of the ASAE 23: 92-98, 103.

11. Phene, C.J. and O.W. Beale. 1976. High-frequency irrigation for water management in humid regions. Soil Sci. Soc. Amer. J. 40: 430-436.

12. Rao, P.S.C., J.M. Davidson and L.C. Hammond. 1976. Estimation of nonreactive and reactive solute front locations in soils. In Residual Management by Land Disposal. Proc. Hazardous Waste Residue Symposium. Tucson, AZ. p. 235-242.

13. Rao, P.S.C., J.M. Davidson and R.E. Jessup. 1981. Simulation of nitrogen behavior in the root zone of cropped land areas receiving organic wastes. In Simulation of Nitrogen Behavior in Soil-Plant Systems (edited by M.J. Frissel and J.A. vanVeen). Pudoc, Wageningen, The Netherlands. p. 81-95.

14. Rawlins, S.L. and P.A.C. Raats. 1975. Prospects for high-frequency irrigation. Science. 188: 604-609.

15. Rhoads, F.M. and R.L. Stanley, Jr. 1973. Response of three corn hybrids to low levels of soil moisture tension in the plow layer. Agron. J. 65: 315-318.

16. Robertson, W.K., L.C. Hammond, G.K. Saxena and H.W. Lundy. 1973. Influence of water management through irrigation and a subsurface asphalt layer on seasonal growth and nutrient uptake of corn. Agron. J. 65: 866-868.

17. Sheridan, J.M., W.G. Knisel, Jr., T.K. Woody, III and L.E. Asmussen. 1979. Seasonal variation in rainfall and rainfall-deficient periods in the Southern Coastal Plain and Flatwoods regions of Georgia. Georgia Agr. Exp. Sta. Res. Bull. 243. 73 p.

18. Skogerboe, G.V., J.W.H. Barrett, B.J. Treat and D.B. McWhorter. 1979. Potential effects of irrigation practices on crop yields in Grand Valley. EPA-600/2-79-149. 193 p.

19. Stegman, E.C., J.T. Musick and J.I. Stewart. 1980. Irrigation water management. In Design and operation of farm irrigation systems (edited by M.E. Jensen) ASAE Monograph 3. Amer. Soc. Agric. Eng., St. Joseph, Michigan. p 763-816.

20. Stewart, J.I., R.D. Misra, W.O. Pruitt, and R.M. Hagan. 1975. Irrigating corn and grain sorghum with a deficient water supply. TRANS. of the ASAE 18: 270-280.

21. Tanner, C.B. 1981. Transpiration efficiency of potato. Agron. J. 73: 59-64.

22. van Bavel, C.H.M. and F.J. Verlinden. 1956. Agricultural drought in North Carolina. North Carolina Ag. Exp. Sta. Tech. Bull. 122. June 1956. 60 p.

23. van Bavel, C.H.M. and J.R. Carreker. 1957. Agricultural drought in Georgia. Georgia Ag. Exp. Sta. Tech. Bull. N.S. 15. 41 p.

EVALUATING IRRIGATION STRATEGIES IN SOYBEANS: A SIMULATION APPROACH

W. G. Boggess, J. W. Jones, D. P. Swaney and G. D. Lynne

Irrigation scheduling is commonly defined as determining when to irrigate and how much water to apply. These decisions will vary depending upon the grower's objectives or decision criteria. Alternative decision criteria include: (1) maximum economic return, (2) maximum yield, (3) maximum response per unit of water and (4) maximum utility based on the expected value and associated variance in net returns. In addition, external costs or institutional constraints may impose restrictions on the grower's use of water or energy for irrigation.

The problem of optimally scheduling irrigation applications, regardless of the particular decision criteria, is complicated by a number of factors. These factors include weather uncertainty, the complex nature of the crop's response to irrigation, uncertain soybean and fuel prices, and possible uncertainty regarding the irrigation water supply. A detailed simulation model incorporating soybean crop, soil-water dynamics and economic decision components has been developed by Swaney et al. (1981) to evaluate soybean irrigation strategies.

The model is designed to study irrigation management in two modes. First specific strategies are input to the model in the following form (Swaney et al., 1981).

> The grower will irrigate on any day of the season if the water content in the root zone of the soil drops below some threshold value (% of available water) specified by the user, and if the time between that day and the previous irrigation is greater than or equal to the return time of his irrigation equipment. If these conditions are met, irrigation water is applied in an amount specified by the user at the start of the season.

The second mode of operation for the model is for real-time irrigation decisions. Here the model is used to supply the user with information on the relative advantages of irrigating today (in real-time) versus waiting one or more days. In this paper, we investigate only predetermined seasonal strategies which are evaluated using the first mode of operation.

The purpose of this paper is to use the simulation model to study alternative irrigation strategies for soybeans using a 55 hectare center pivot irrigation system. Specific objectives were to: (1) evaluate alternative season-long, fixed irrigation strategies; (2) evaluate the sensitivity of the maximum net returns strategy to changes in the price of soybeans and the price of fuel; and (3) evaluate the sensitivity of soybeans to water stress imposed in various crop growth stages, as simulated by the crop model.

The authors are: W. G. BOGGESS, Assistant Professor, and G. D. LYNNE, Associate Professor, Food and Resource Economics Department; J. W. JONES, Professor, and D. P. SWANEY, Scientific Programmer, Agricultural Engineering Department, University of Florida.

ECONOMIC DECISION FRAMEWORK

Most decision problems consist of three components: (1) an objective or decision criterion; (2) a set of alternative choices; and (3) a set of costs and constraints which limit the choice set. These components are recognized in the economic framework developed below. The formulation recognizes that the irrigator not only has an economic demand for water, but that he must also generally operate his own water supply service with its concomitant investment and operation costs. The formulation describes, conceptually, the demand and supply relationships for irrigation water, and broadly outlines the decision calculus of the manager who wishes to maximize his net returns (Lynne and Carriker, 1979).

In the short run yield can be expressed functionally as:

$$Y = f\,(\overline{W},\overline{I}) \qquad \text{Eq. (1)}$$

where $\overline{W}$ is the vector of seasonal weather, $\overline{I}$ is the vector of seasonal irrigation amounts, and Y is the resulting yield. It is assumed for the strategies evaluated in this paper that the levels of other production inputs remain invariant. The irrigator's short run water demand function is derived from the short run production function and the expected price of soybeans, and is expressed functionally as

$$I^d = g(r_d, \overline{W}, p) \qquad \text{Eq. (2)}$$

where r_d is the marginal value of irrigation water, W is the vector of seasonal weather, p is the expected price of soybeans, and I^d is the seasonal demand for irrigation water.

The firm's irrigation water supply function can be expressed as:

$$I^s = h\,(A, L, r_i, \ldots, r_h) \qquad \text{Eq. (3)}$$

where the functional form of h depends upon the type of irrigation system, A is a water availability factor, L represents limitations, conditions or other regulations imposed by public institutions on private irrigation decisions, $r_i, \ldots, r_h$ are the prices of inputs (e.g. fuel and labor) used to produce irrigation water, and I^s is the vector of irrigation water supplied. For this analysis, irrigation costs include only operating costs assuming the grower already has center pivot irrigation system. In the short run the water supply function is

$$I^s = h(c\,|\,A,L) \qquad \text{Eq. (4)}$$

where c is the marginal costs of irrigation water and the other variables are defined as before and are assumed invariant during the time period to which the analysis is applied.

The actual quantity of water applied by an irrigator seeking to maximize his short run net returns will be the quantity which simultaneously satisfies equations 2 and 4. The ultimate demand for water by the firm is a function of the biological plant-water relationships, input costs, soybean price, weather, and any institutional restrictions imposed on irrigators.

MODEL AND METHODS

The optimization technique utilizes a simulation model of the soybean crop (Wilkerson et al., 1981) which is sensitive to photosynthetically active radiation (PAR), daily temperature, and soil water stress which is determined by a soil water balance model (Jones and Smajstrla, 1979) to simulate the production surface functionally represented in equation 1. The crop growth and soil water models are controlled by a set of subroutines which include user interfaces to specify and update model parameters, a multi-year driver, which runs the simulation under 17 seasons of historical weather in order to obtain average values of the decision criteria, and an economic and statistical subroutine which accumulates the results from different weather seasons and calculates the decision outcomes. The model components are described in more detail in Swaney et al. (1981).

The results suggested by this analysis can not be extrapolated to other locations, crops, irrigation systems or soil types. The current version of the model is restricted to a deep, well drained sandy soil typical of Florida conditions. Also, the 17 years of historical weather data were from central Florida, a sub tropical area. However, the approach used in this analysis could be applied to other crop systems.

Optimization of Preseason Strategy with Constant Parameters

In order to determine the best parameters of the irrigation scheduling strategy for the various decision criteria, parameters were varied giving a series of simulation runs. These parameters were depth per application (i_j) and soil moisture irrigation threshold (τ_j) for the center pivot system.

The soil moisture threshold (τ_j) represents the ratio of water actually available in the root zone divided by water available in the root zone when the soil is at field capacity. For the sandy soils in the study area, 7.5 cm. of water is available in the 125 cm. mature root zone when the soil is at field capacity. An irrigation threshold of 70 percent indicates that irrigation is initiated when 30 percent or 2.25 cm. of water have been depleted from the root zone. An irrigation efficiency of 0.75 was used in this analysis.

The optimal values of the parameters can be obtained for a finite set by simply evaluating the relevant objective function for each pair and selecting the pair that results in the optimum value of that function. For this evaluation, the set of i's selected were the integer values 0 to 7 centimeters, inclusive. The set of τ's ranged from 0 to 100 percent in steps of 10 percent. Thus a total of 88 pairs were simulated for the 17 seasons of historical weather. The expected value and standard deviation of yield, net returns, irrigation water applied, and energy use were calculated for each of the runs.

Sensitivity Analysis of Maximum Net Returns Strategy to Changes in Soybean and Fuel Prices

Equations 2 and 4 indicate that the quantity of irrigation water which maximizes net returns depends upon the price of soybeans and the price of fuel, the latter having its impact on the marginal costs of irrigation. In order to evaluate the sensitivity of the model to changes in these prices, the price of soybeans was parameterized from \$.08 per kilogram to \$2.32 per kilogram. Likewise the price of diesel fuel was parameterized from \$.20 per liter to \$1.60 per liter. The marginal costs of irrigation water applied with a center pivot system were calculated using an irrigation cost simulator (d'Almada et al., 1981). The net returns for each of the 88 simulation runs were then compared.

Evaluation of the Sensitivity of the Model to Water Stresses Imposed in Different Growth Stages

It is commonly recognized that crop response to (and thus the value of) irrigation water varies depending upon the growth stage of the crop. To evaluate the sensitivity of the soybean model to water stresses occurring in different stages a series of simulation runs were made varying the level of imposed stress in four different two-week periods. The four periods were specified to occur prior to R2, between R2 and R4, between R4 and the end of flowering, and between the end of flowering and R7. Growth stages R2, R4, and R7 are defined by Fehr et al. (1971). (See Wilkerson et al. (1981) for a more complete description of these periods.) Period one begins on the 19th day after planting and ends on day 32; period two begins on day 49 and ends on day 62; period three begins on day 74 and ends on day 87; and period four begins on day 105 and ends on day 118. The soil water content in the root zone varied from 80 percent to 30 percent of available water in steps of 10 percent in each of the four periods such that all combinations were simulated. On days outside of the four stress periods the soil water content was set to its field capacity.

RESULTS

Preseason Irrigation Strategy Optimization

The seasonal irrigation strategy simulations produced a net returns surface as a function of irrigation threshold (τ) and irrigation amount per application (i) (Fig. 1). Maximum net returns occured at a value of 70 percent of field capacity as the threshold and a 1 cm. application per irrigation. This combination yields an expected net return of approximately $553 per hectare compared to $344 for the no irrigation case and -$10 for the minimum profit strategy of applying 1 cm. of water whenever soil water drops below its water holding capacity.

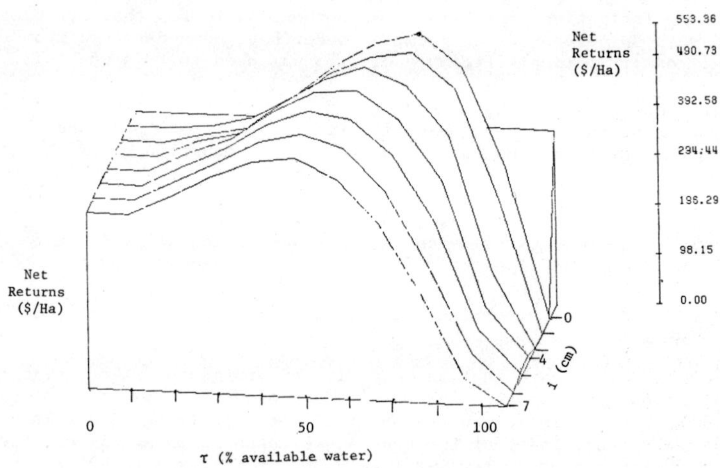

Fig. 1. Plot of profit versus irrigation threshold (τ) and amount of water applied per application (i).

The (simulated) crop experiences water stress whenever soil water drops below 80 percent of available water. Thus, the maximum net returns strategy allows some stress to the crop, since the marginal cost of irrigation exceeds the marginal value of the increase in yield toward no stress. At the 70 percent threshold, 2.25 cm. of the 7.5 cm. maximum available soil moisture have been depleted. The optimal strategy calls for less than half of this deficit to replaced suggesting that given the frequency and magnitude of rainfall in the study area, that some storage capacity should be left for storage of rain water.

The maximum yield objective results in optimal values of 1 cm. per application and the 90 percent threshold. At this combination the expected yield per hectare is 123 kg. greater than the maximum net returns strategy, but expected net returns per hectare are \$258 lower.

A third possible criterion for scheduling irrigation is to maximize yield response per unit of irrigation water. Irrigators using this criterion should choose the 50 percent threshold and 2 cm. per application strategy. This strategy uses on average only 40 percent as much irrigation water as the maximum net returns strategy. However, the expected yield is approximately 506 kg. lower and expected net returns \$50 lower than the maximum net returns situation.

One final objective that irrigators might use to make irrigation scheduling decisions is to maximize utility, where utility is a function of the expected value $E(\pi)$ and associated variance $V(\pi)$ in net returns:

$$U = U\,[E(\pi),\ V(\pi)] \qquad \text{Eq. (5)}$$

The specific functional form of this relationship varies by individual depending on each individual's pyschological aversion to risk. For risk averse individuals the following conditions hold:

$$\frac{\delta U}{\delta E(\pi)} > 0 \text{ and } \frac{\delta U}{\delta V(\pi)} < 0 \qquad \text{Eq. (6)}$$

This indicates that a risk averse individual faced with two strategies of equal expected value will choose the one with the smaller variance. Likewise, the choice between two strategies of equal variance will be based on the higher expected value. However, the choice between two strategies where the first has both a greater expected value and variance than the second can be made only with knowledge of the specific functional form of the individual's utility function.

The expected value of the net returns has been plotted against the associated standard deviation of net returns for several of the simulated strategies (Fig. 2). Note that the maximum net returns strategy of 70 percent and 1 cm. has both a higher expected value and smaller standard deviation and thus dominates all strategies with lower thresholds. However, as the threshold level is increased from 70 percent, both the expected value and the standard deviation decrease. Thus, a risk averse individual may choose to irrigate at a higher threshold and thus forego some expected net returns in return for reduced variability.

Price Sensitivity Results

The 70 percent 1 cm. strategy maximizes expected net returns when the price of soybeans is \$.2572 kg. and the price of diesel fuel is \$.32 per liter. However, as equations 2 and 4 indicate the optimal strategy may change in response to changes in either of these prices. Tables 1 and 2 report the results of the price sensitivity analysis.

Table 1 indicate that for soybean prices of \$.08 per kg. or lower it does not pay to irrigate. As the price of soybeans increases the optimal irrigation threshold value increases. The previous strategy of 70 percent and 1 cm. is optimal for soybean prices between \$.18 and \$.73 per kg. Notice that the price of soybeans must increase to \$2.34 per kg. before the maximum yield strategy of 90 percent and 1 cm. also maximizes expected net returns.

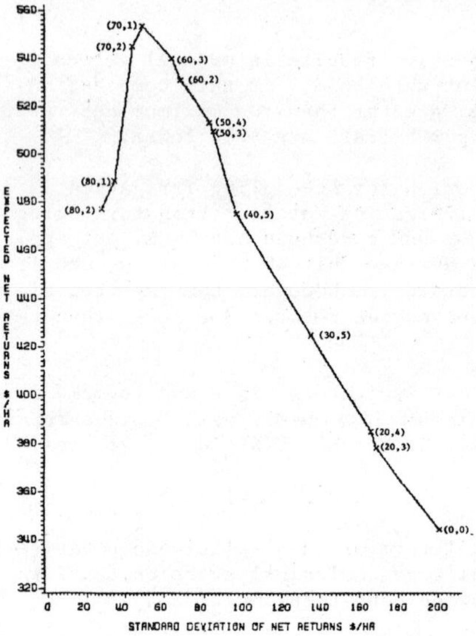

Fig. 2. Plot of expected value versus standard deviation of net returns for various combinations of irrigation threshold and depth per application (τ,i).

Similar results occur with respect to changes in fuel prices except that an inverse relationship exists between the optimal threshold and fuel prices (Table 2). The previous strategy of 70 percent and 1 cm. remains optimal for diesel fuel prices between $.80 and $3.00 per gallon. For prices of $6.00 per gallon or higher it doesn't pay to irrigate. The price of diesel fuel must drop to zero before it pays to increase the threshold from 70 percent to 80 percent.

Notice that the optimal net returns strategy is bounded from below by the 50 percent and 2 cm. strategy that maximizes yield per unit of water and bounded from above by the maximum yield strategy of 90 percent and 1 cm. The maximum net returns strategy will generally lie somewhere between these two bounds (or call for zero irrigation) depending upon the relative prices of fuel and soybeans.

The key factor behind the sensitivity of the maximum net returns strategy to price changes is the ratio of soybean price to the marginal cost of water. As this ratio increases, the optimal quantity of irrigation water increases (Table 1) as the ratio declines (Table 2), the converse holds. Since energy used for irrigation is a direct function of the amount of water applied, restrictions on either water or energy use will have identical impacts on the optimal strategy.

Growth Stage Sensitivity Results

The truly optimum irrigation strategy will vary by growth stages if the crop's response to water varies by stage. These differential effects must be accounted for in order to insure a true maximum profit strategy. Said somewhat differently, equations 2 and 4 must be satisfied within each growth stage. Preliminary estimates suggest the true optimum is somewhat different than the 70 percent and 1 cm. fixed strategy for the entire season.

Table 1. Sensitivity Analysis of the Maximum Net Returns Strategy to Changes in Soybean prices, holding the Marginal Cost of Water Constant at $6.07/cm.

	Units $/kg.	Soybean Prices .0800	.1470	.1837	.2572	.3674	.4409	.7347	2.320
Expected Net Returns[a]	S/Ha.	-170.01	94.50	243.79	552.73	1015.93	1324.87	2560.64	9219.29
Expected Yield	Kg/Ha.	2915.81	3953.96	4203.27	4203.27	4203.27	4203.27	4203.27	4326.43
Irrigation Energy Use	MCAL's	0.0	8947.15	13383.76	13383.76	13383.76	13383.76	13383.76	45105.55
Irrigation Water Use	Cm	0.0	14.24	21.29	21.29	21.29	21.29	21.29	71.76
Water Applied Per Application	Cm	0.0	2.0	1.0	1.0	1.0	1.0	1.0	1.0
Soil Water Threshold	%	0	60	70	70	70	70	70	90

[a]Returns are net above variable operating costs. Irrigation operating costs are based on a 55-hectare center pivot system.

Table 2. Sensitivity Analysis of the Maximum Net Returns Strategy to Changes in Fuel Prices, Soybean Price Constant at $.2572/kg.

	Unit S/cm	Marginal Cost of Water[a] 1.54	4.34	5.74	6.79	8.54	12.04	19.04	22.71
Expected Net Returns[b]	S/Ha	687.17	596.55	552.57	519.58	464.61	377.22	242.86	171.76
Expected Yield	Kg/Ha	4309.77	4203.27	4203.27	4203.27	4203.27	3946.96	3697.73	2915.81
Irrigation Energy Use	MCAL	23477.05	13383.76	13383.76	13383.76	13383.76	8762.30	5545.75	0.0
Irrigation Water Use	Cm	37.35	21.29	21.29	21.29	21.29	13.94	8.82	0.0
Water Applied Per Application	Cm/Ha	1.0	1.0	1.0	1.0	1.0	1.0	2.0	0.0
Soil Water Threshold	%	80	70	70	70	70	60	50	0

[a]The marginal cost of irrigation water is directly related to the price of fuel used to power the System. The marginal costs reported in the table correspond to diesel fuel prices of $ .00, .80, 1.20, 1.50, 2.00, 3.00, 5.00, $6.05/gal. respectively, from let to right

[b]Returns are net above variable operating costs. Irrigation operating costs are based on a 55-hectare center pivot system.

The growth stage water stress simulations produce a series of yields and associated total seasonal transpiration (Fig. 3). In order to condense the information contained in Fig. 3, the following regression equation was estimated:

$$Y = T_1^{0.698} \; T_2^{0.196} \; T_3^{1.034} \; T_4^{0.690} \qquad R^2 = 0.972 \qquad \text{Eq. (7)}$$
$$(5.65) \; (4.91) \; (20.14) (11.54)$$

where Y is the ratio of actual yield divided by potential yield, and T_i is the ratio of actual transpiration in growth period i divided by potential transpiration in growth period i. The values in parentheses represent the calculated t statistic for each parameter estimate. All parameters are significant at the .0001 level.

Equation 7 indicates that soybeans are most sensitive to water stress in period 3 and least sensitive in period 2. These results suggest that a truly optimal irrigation strategy would call for varying threshold levels depending upon the growth stage of the crop with a relatively high threshold in period 3, a relatively low threshold in period 2 and an intermediate threshold in periods 1 and 4.

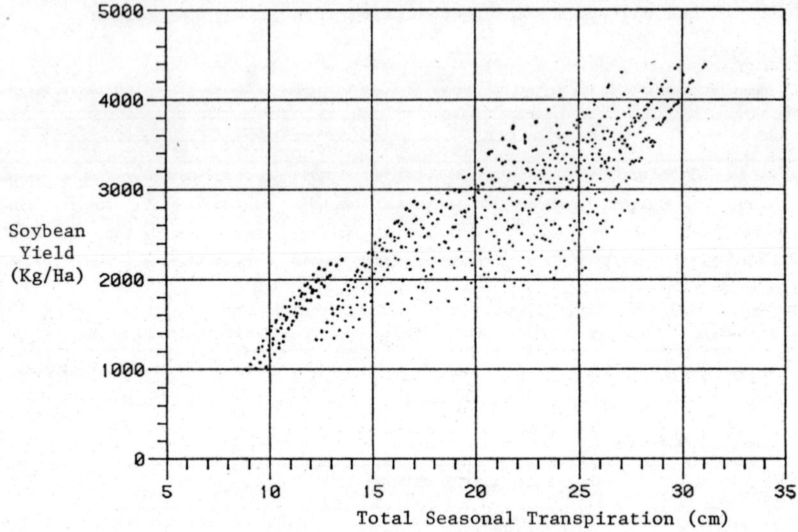

Fig. 3. Plot of simulated yield versus total seasonal transpiration for various combinations of water stresses imposed by growth stage.

SUMMARY

A simulation model was used to evaluate preseason, constant parameter irrigation strategies. Optimal strategies were defined for alternative decision criteria including: (1) maximum economic returns, (2) maximum yield, (3) maximum response per unit of water and (4) maximum utility based on the expected value and variance in net returns. The sensitivity of the maximum net returns strategy to changes in soybean and fuel prices was analyzed. Finally, the sensitivity of the soybean crop model to water stress events imposed in different crop growth stages was studied.

CONCLUSIONS

The model results suggested that the irrigation strategy using a 70 percent threshold and 1 cm. per application maximizes net returns for soybeans under center pivot irrigation in Florida. This strategy is relatively robust with respect to changes in both soybean and fuel prices. The maximum net returns strategy is bounded from below by the maximum yield per unit of water strategy and bounded from above by the maximum yield strategy. A risk averse grower would apply the maximum net returns strategy as a minimum since it dominates all lower thresholds, and depending upon his degree of risk aversion, might prefer a higher threshold strategy.

The analysis has not progressed far enough at this time to specify an optimal irrigation strategy by growth stage. In addition, more data are needed to evaluate the ability of the model to accurately simulate the effects of stress in different growth periods. Work is currently underway to compare the model with results from several field irrigation experiments (Hammond, L.C., 1981). Then efforts will be made to determine the optimal thresholds and application rates for each growth stage and to evaluate the sensitivity of these parameters to changes in prices. It is clear, however, that soybean crop response to water stress varies by growth stage and that by tying the irrigation scheduling strategy to the growth stage of the crop we can improve profits, as compared to the preseason constant parameter strategy.

REFERENCES

1. d'Almada, P., G. D. Lynne, and A. Smajstrla. Irrigation cost program users reference manual. Unpublished Report, Food and Resource Economics Department, University of Florida, Gainesville.

2. Fehr, W. R., C. E. Caviness, D. T. Burmood, and J. S. Pennington 1971. Stage of development descriptions for soybeans, _Glycine_ _Max_ (L.) Merrill. Crop Science 11:929-931.

3. Hammond, L. C. 1981. Unpublished soybean irrigation experimental data. Soil Science Department, University of Florida, Gainesville, FL.

4. Jones, J. W. and A. Smajstrla. 1980. Application of modeling to irrigation management of soybeans. In World Soybean Research Conference II Proceedings, F. T. Corbin (ed.). Westview Press, Boulder, CO. pp. 571-599.

5. Lynne, Gary D. and R. Carriker. 1979. Crops response information for water institutions, American Society of Civil Engineers, J. Irrigation and Drainage Division 105:IR3.

6. Swaney, D. P., J. W. Jones, W. G. Boggess, G. G. Wilkerson, and J. W. Mishoe. 1981. A crop simulation model for evaluating within season irrigation decisions. American Society of Agricultural Engineers Paper No. 81-4015, St. Joseph, MI.

7. Wilkerson, G. G., J. W. Jones, K. J. Boote, K. I. Ingram, and J. W. Mishoe. 1981. Modeling soybean growth for crop management. American Society of Agricultural Engineers Paper No. 81-404. St. Joseph, MI.

OPTIMIZING IRRIGATION MANAGEMENT OF TRAVELING GUNS

Richard A. Schoney   Leonard R. Massie   Ted F. Bay
Member ASAE

A major irrigation management goal is decreased water consumption and energy costs while maintaining high yields. In the humid Midwest weather is particularly uncertain with precipitation being unpredictable and daily evapo-transpiration rates highly variable. This is not a significant problem when "turn-around" times are three days or less as with the modern electric center pivots, however, this can be a major problem in the management of large volume traveling guns. In a risky and uncertain environment, good irrigation management typically dictates a "safety first" strategy of high water application rates. This decision rule is based on the premise that cost of under application (reduced yields) is greater than the cost of over application (variable cost of irrigation and traveling gun). The problem is further complicated because the irrigated acreage is a decision variable.

The determination of the optimal irrigation management and irrigated acreage is complicated by (1) system management limitations, (2) moisture stress-yield relationships, and (3) decision-maker aversion to risk. This study reports the simulated trade-off in choice of watering strategies, irrigated acreage, net returns and risk (as measured by variability of returns) for a typical traveling gun irrigator in the humid Midwest.

APPROACH

The approach can be divided into these steps. Moisture stress in corn was simulated. Weekly indices representing lack of moisture stress (1-stress coefficient) were estimated. These indices were included as independent variables along with fertilizer variables in estimating a statistical yield response function.

The moisture stress simulator and yield response model were combined with an irrigation management model which included the system use limitations and irrigation decision rules. The combined simulation model was used to evaluate alternative decision rules.

The authors are: R.A. SCHONEY, Assistant Professor, Department of Agricultural Economics, L.R. MASSIE, Associate Professor, Department of Agricultural Engineering, T.F. BAY, Specialist, Department of Agricultural Economics, University of Wisconsin-Madison.

Research supported by the College of Agriculture and Life Sciences, University of Wisconsin-Madison and by USDA SEA Grant Formula, Project No. 5127.

Corn Moisture Stress Model

The corn moisture stress model was based on a four segment soil profile with moisture moving from segment to segment. The model has five important submodels: (1) soil moisture balance, (2) corn phenological development, (3) soil moisture extraction, (4) potential ET and (5) stress. The soil moisture balance submodel was based on balancing each of four segments within the profile. Components of the soil moisture balance include: precipitation, upward moisture movement, actual evapo-transpiration and downward moisture movement.

The development of the corn leaf conopy, root structure and plant is a function of temperature degree days (base 50). Potential moisture extraction for each of the four segments is based on a cumulative moisture extracting pattern from Borg (P. 55, 1980):

$$CUM = 9.3e^{0.0229PERROOT} \quad (1)$$

CUM = cumulative moisture extraction (%)

PERROOT = effective root depth as a percent of maximum rooting depth

The extraction pattern of Eq. 1 is modified when high amounts of precipitation make the moisture supplying ability of the top layer greatly exceed the bottom layers.

The key algorithm in the model is a potential evapo-transpiration (PET) subroutine which is based on solar radiation and developed by C.B. Tanner (1976).

Potential ET (PET) consists of evaporation (E) and transpiration (T) which are modified by leaf area index (which varies between 0 and 1):

$$PET = LAI*T + (1-LAI)*E \quad (2)$$

Actual ET is based on the relationship between available soil moisture (θ_s) and a critical soil moisture level (θ_c), both in percentage terms:

$$AET = PET \text{ if } \theta_s > \theta_c \quad (3)$$

$$AET = \frac{\theta_s}{\theta_c} * PET$$

The critical soil moisture level is a threshold where crop stress starts and is variable depending upon PET (Borg, 1980):

$$\theta_c = 20.9 + 28.1\, PET \quad (4)$$

where PET is in cm and θ_c is in percentage terms.

The degree of moisture stress (s) is defined as:

$$s = 1-AET/PET \quad (5)$$

Yield Stress/Fertility Model

Yield reduction due to untimely or improper irrigation management is based on a nonirrigated yield stress model. The weekly yield multiplier indices (YMI) are equal to 1.0 - s. Various combinations of multiplying the weekly YMI were tried with mixed results. The final YMI was estimated as the product of nine weekly YMI variables starting with the week of first tassel. The regression analysis is based on fertility trials on a Plano silt loam for the years 1977-1980 conducted by Walsh, Moncrief and Schulte at the Arlington Experimental Station. The Plano silt loam has excellent water holding capabilities making it somewhat undesirable for a stress model.

The fertility variables were estimated based on the YMI, available nitrogen (N) and potassium (K). Available nitrogen is based on applied nitrogen fertilizer combined with an estimate of the nitrogen available from soil organic matter. It was estimated that approximately 5 percent of the soil organic matter is organic nitrogen, with approximately 2 percent of the organic nitrogen becoming available during the growing season. Available potassium is based on soil test potassium combined with an estimate of the potassium available from applied potassium fertilizer. Based on prior experience at the Arlington test site, it was estimated that approximately 1 pound in 6 of the applied potassium fertilizer became available during the growing season. The estimated response model is based on the translog (Schoney, et al. 1981) where yield (Y) is in metric tonnes per hectare and YMI ranges between 0 and 1.

$$\begin{aligned} \ln Y = & \underset{(-9.37)}{-9.703} + \underset{(10.46)}{4.1977\ln N} + \underset{(3.07)}{.2446\ln K} - \underset{(-9.49)}{.2660\,(\ln N)^2} - \underset{(-5.70)}{.04816\,(\ln K)^2} \\ & - \underset{(-1.08)}{.01471\,(\ln N)(\ln K)} + \underset{(1.56)}{.111\ln YMI} \end{aligned} \qquad (6)$$

R^2 = 82.4

s = .1385

t values are in parentheses

Irrigation Management Model

The Traveling Large-volume Sprinkler (Gun) gives the irrigator considerable flexibility in setting the irrigated acreage. The flexibility, however, comes at a relatively high labor and management requirement when compared to the center pivot systems. In general, many irrigators try to cut fixed costs by irrigating as much acreage as possible per traveler, generating relatively long turn-around times. Increasing the turn-around times increases the difficulty of management by (1) extending the period of projected water balances and (2) expanding the number of individual water balances.

In the irrigation management model key decision variables are (1) irrigated acreage and associated turn around time in days (2) soil moisture at which irrigation is initiated (θ_i) and (3) soil moisture at which irrigation is terminated. The irrigated acreage is divided into units according to the number of days required to complete the full irrigation cycle (turn-around time). Each unit, the land irrigated in one day representing 2 or 3 moves of the traveler, carries its own soil moisture balance, estimated stress and yield. Thus a five day turn-around requires five individual units, each treated separately. Irrigation is initiated when the soil moisture of the last field in the cycle is projected to fall below soil moisture level θ_i, before completing the irrigation cycle. Irrigation is terminated when the soil moisture rises above θ_t. Projections are based on current soil moisture levels, a weighted average of the past ET and a rain/no rain variable.

RESULTS

The yields and net returns were simulated for the years 1963-1980 based on weather data from the Experimental Station at Arlington, Wisconsin and the Lapeer soil series. The Lapeer is a medium-textured soil of sandy loams with a relatively high clay content in the upper profile. The basic profile and moisture characteristics of the soil are given in Table 1.

Table 1. Estimates of Significant Soil Properties

Depth From Surface (cm)	Dominant USDA Texture	Permeability (cm/hr)	Available Water Capacity (cm/cm)
15	Fine Sandy Loam	1.5 - 5.1	0.16 - 0.18
15 - 79	Heavy Sandy Loam[a/]	1.5 - 5.1	0.15 - 0.17
79 - 152	Gravelly Sandy Loam	5.1 - 15.2	0.07 - 0.09

[a/] The weighted average texture is heavy sandy loam, but individual subhorizons are light sandy clay loam or sandy loam. The average clay content ranges from about 15 to 18%.

Net returns for irrigated production are based on a traveling gun irrigator with the following investment requirements:

Well	$ 6,000
Pipe (15cm main)	5,640
Traveler and Hose	15,200
Pump, Motor, and Controls	12,000
Electrical Line and Hookup	1,650
Total Investment	$40,490

The annual capital recovery charge and estimated repair costs totaled $8,524.

Economic feasibility (profitability) of irrigation is based on a comparison of dryland and irrigated returns. Yields were estimated with 5 to 10 day turn-around times and the soil moisture threshold (θ_1) varied from 35% to 50% of available moisture. Potential dryland yield was estimated to be 7.5 metric t/ha with potential irrigated yield of 9.4 t/ha. Water application was allowed to vary between 1.3 cm and 2.5 cm at each application.

Corn is priced at $51.83/t with harvest and drying costs of $11.16/t leaving a net return of $40.67/t. The variable costs/ha of irrigation consist of $6.93/100m^3 of water applied and a labor charge of $1.20/move. Estimated yields, water application, and net returns for the 5 and 10 day turn-arounds are shown in Table 2. Irrigated returns are based on 20.24 ha irrigated/gun with a 5 day turn-around and 40.47 ha irrigated/gun with a 10 day turn-around, both schedules based on 2 moves/day.

Of the 18 years used in this study only 3 (1967, 1979, and 1978) show any significant yield reduction when going from a 5 to 10 day turn-around. When a 5 (10) day turn-around period is used the simulation model was required to project weather 4 (9) days in advance so as to "anticipate" the soil moisture on the last plot to be irrigated. Wisconsin seldom experiences hot, dry periods of longer than 2 weeks. For example in only 2 years out of 18 years was there less then 5.08 cm of rainfall in August. The combination of humid weather and the heavier soils of the Midwest such as the

Table 2: Irrigated Production and Returns (θ_i = 50%)

Year	Est. Dryland Prod.	Net Returns	Yield 5 Day Turn-around	Yield 10 Day Turn-around	Water Applied 5 Day Turn-around	Water Applied 10 Day Turn-around	Moves 5 Day Turn-around	Moves 10 Day Turn-around	Net Return 5 Day Turn-around	Net Return 10 Day Turn-around
	(t/ha)	($/ha)	(t/ha)		(m^3/ha)		(# of moves/ha)		($/ha)	
1963	5.7	576	9.4	9.4	1524	1549	6	7	$465	$670
1964	5.9	591	9.4	9.4	1626	1295	7	6	455	677
1965	6.9	694	9.4	9.4	1270	1372	5	6	474	680
1966	7.0	707	9.4	9.4	965	1118	4	5	484	689
1967	5.2	526	9.4	8.8	1778	1219	7	5	452	625
1968	6.6	660	9.4	9.4	1270	1143	5	5	474	687
1969	5.3	536	9.0	8.2	1118	686	5	3	440	583
1970	5.2	519	9.4	9.2	2261	1651	9	7	435	645
1971	5.5	553	9.3	9.2	1651	1499	7	7	445	652
1972	7.5	754	9.4	9.4	737	787	3	4	494	702
1973	6.3	633	9.4	9.4	1524	1473	6	6	465	675
1974	5.7	571	9.3	9.3	1372	1397	6	6	460	672
1975	6.1	613	9.4	9.3	2032	1473	8	6	442	665
1976	5.6	563	9.4	9.4	2134	1854	9	8	437	655
1977	6.4	642	9.4	9.4	1270	1219	5	6	474	682
1978	4.9	494	9.2	8.4	1778	1219	7	5	430	586
1979	7.5	756	9.4	9.4	508	762	2	4	504	702
1980	6.6	660	9.4	9.4	1499	1194	6	5	465	687

Lapeer used in this study allows an irrigator to extend an irrigation system further than usually practiced in the West. The longer turn-around time requires greater management with careful attention given to soil water capacity and current moisture level, ET, and irrigation strategy.

The 18 year averages in Table 3 further show the result of stretching a system while also anticipating weather conditions. Stretching irrigation turn-around from 5 to 10 days resulted in a yield loss of .1-.2 tonne/ha. Starting irrigation with an anticipated θ_i = 50% of soil capacity rather than θ_i = 35% resulted in a .1-.2 tonne yield advantage (10 and 5 day turn-around, respectively) for θ_i = 50%. The yield increases required approximately 6.3 cm more water for the 5 day turn-around cycle and 1.9 cm for the 10 day turn-around. With a \$.07/m^3 cost of water applied the increased yield easily pays for the increased water requirement in both cases.

Table 3: Average Yield, Water Application, and Net Returns (1963-1980)

		Yield	Water Applied	Net Returns
		(tonne/ha)	(m^3/ha)	(\$/ha)
Dryland		6.1	--	613
Irrigated:	Turn-around			
θ_i = 35%	5 days	9.2	1143	460
	10 days	9.1	1194	660
θ_i = 50%	5 days	9.4	1473	460
	10 days	9.2	1270	665

The main point brought out by Tables 2 and 3 is that on a medium-textured soil, the watering schedule can be extended from a 5 to 10 day turn-around without a significant yield reduction. The 1 or 2 dry years experienced out of 10 may significantly reduce yields from the overall average but is very unlikely to result in a complete failure. The results of Tables 2 and 3 are based on using 2 moves per day for both the 5 and 10 day turn-around. If the availability of labor allows 3 moves per day, fixed costs can be reduced further.

REFERENCES

1. Black, T.A., W.R. Gardner and G.W. Thurtell. 1969. The Production of Evaporation, Drainage, and Soil Water Storage for a Bare Soil. Soil Science America Proc. 33:655-660.

2. Borg, Heinz. 1980. Plant Available Water. M.S. Thesis. University of Wisconsin-Madison, Madison, WI.

3. Denmead, O.T., and R.H. Shaw. 1960. The Effects of Soil Moisture Stress at Different Stages of Growth on the Development and Yield of Corn. Agronomy Journal 52(5):272-274.

4. Dylla, A.S., H. Shull and D.R. Timmons. 1980. Computer Simulation Sprinkler Irrigation Practices in Minnesota. Transactions of the ASAE. 23(6):1419-1423.

5. James, L.G. and C.L. Larson. 1976. Modeling Infiltration and Redistribution of Soil Water During Intermittent Application. Transactions of the ASAE. 19(3):482-488.

6. Jensen, M.R., H.R. Haise, and J.L. Wright. 1978. The Role of Evapotranspiration Models in Irrigation Scheduling. Transactions of the ASAE. 21(1):82-87.

7. Mitchel, Michael J., 1978. Soil Survey of Columbia County, Wisconsin. USDA-Soil Conservation Service.

8. Moore, C.V. 1961. A General Analytical Framework for Estimating the Production Function for Crops Using Irrigation Water. American Journal of Agricultural Economics. 43:876-888.

9. Morey, R.V., J.R. Gilley, F.G. Bergsrud and L.R. Dinkzwager. 1981. Yield Response of Corn Related to Soil Moisture. Transactions of the ASAE. 23(5):1165-1170.

10. Musick, J.T. and D.A. Dusek. 1980. Irrigated Corn Yield Response to Water. Transactions of the ASAE. 23(1):92-98.

11. Schoney, R.A., T.F. Bay, and J.F. Moncrief. 1981. Use of Computer Graphics in the Development and Evaluation of Response Surfaces. Agronomy Journal. 73(3):437-442.

12. Sylla, Dean S. 1980. A Computer Soil Moisture Model for Making Irrigation Decisions: Potatoes on Plainfield Loamy Sand. M.S. Thesis. University of Wisconsin-Madison, Madison, WI.

13. Tanner, C.B., and W.A. Jury. 1976. Estimating Evaporation and Transpiration from a Row Crop During Incomplete Cover. Agronomy Journal 68:239-243.

SCHEDULING FOR OPTIMUM WATER USE

by Marshall English, Michael Glenn and John VanSickle*

Current irrigation scheduling practice is generally designed to maximize application efficiencies without adversely affecting crop yields. However with the high cost of energy and increasing competition for water irrigation scheduling may be gradually reoriented toward a new goal, economically optimum water use. This paper outlines a proposed approach to optimum scheduling. The approach involves the use of a crop stress indicator as an index of water requirements, combined with the use of mathematical filtering techniques to detect critical stress values.

Irrigation scheduling, as it is now generally practiced, involves irrigating before soil moisture reaches a critical point at which crop yields will be adversely affected. The goal has been to use water as efficiently as possible to achieve maximum yields. However, the approach usually taken in practice is to schedule irrigations conservatively. Farmers often apply water earlier and in greater quantities than recommended (English et al., 1980; Brost, 1977). Scheduling services tend to be conservative in their assessments of allowable depletion. For example irrigations may be scheduled in such a way as to refill the profile in those areas of a field in which water-use is highest (Gear et al., 1976). While this practice results in overestimation of field-average moisture depletion it is consistent with the fundamental constraint of maintaining full yields.

Optimal scheduling will need to be more precise. Rather than calling for an irrigation when moisture depletion is approaching a critical point, the goal will be to recognize when it has reached a critical point. In addition the critical point may not be that point at which yields begin to be affected, but rather some later point at which yields are reduced by a specified amount.

Optimum Irrigation Practices

Economically optimum irrigation practice involves reducing water use from the yield maximizing level to the level at which marginal cost equals the value of the marginal product. Such a strategy of deliberate under-irrigation is sometimes referred to as deficit irrigation. Broadly speaking, there are two ways to practice deficit irrigation. One is to apply water frequently but in amounts insufficient to refill the soil profile (high frequency deficit irrigation). The other is to prolong the intervals between irrigations until actual evapotranspiration has fallen well behind the maximum rate, and then apply sufficient water to refill the soil profile (low frequency deficit irrigation). It has been claimed that the first of these techniques may result in somewhat higher yields (Miller, 1977), but that claim is still in doubt (Ferreres et al., 1978). Under certain circumstances the first approach may reduce capital costs (Rawlins and Raats, 1975), while the second approach may reduce capital, maintenance and labor

* Assistant Professor, Department of Agricultural Engineering; Assistant Professor, Department of Crop Science; and Assistant Professor, Department of Electrical and Computer Engineering, Oregon State University.

costs (English and Nuss, 1980).

It is useful to consider the relevant characteristics of an optimally irrigated field. The most obvious characteristics will be a dry soil profile at the time of irrigation. There will be no appreciable reduction in water use until soil moisture declines to a point at which evapotranspiration is significantly inhibited. That point varies with crop, soil type and weather conditions, but as a rough rule of thumb it can be said to occur at a soil matric potential of between -0.5 and -1.0 bars (Feddes, et al., 1978, p. 19).

Soil moisture may also be more variable under deficit irrigation. Application efficiency may be increased as irrigation adequacy is reduced (Shearer, 1978). So deliberately operating an irrigation system to achieve low adequacy can be to the farmer's advantage (English and Nuss, 1980). Soil moisture variability may be aggravated if irrigation adequacy is reduced.

Crop growth and development may also be less uniform in an optimally irrigated field. If the distribution system has been designed for low frequency deficit irrigation the return intervals between irrigations will be long (English and Nuss, 1980) and some parts of the field will receive water at more favorable times than others. The result may be more variable crop conditions.

Based on these premises, one can speculate on the directions in which irrigation scheduling is likely to develop. The discussion which follows deals first with various indicators which might be used to determine irrigation requirements. Secondly, the use of filtering to enhance the effectiveness of these indicators is proposed.

1. Indicators of crop Water Requirements

An indicator of crop water requirements should be related to crop yields, since yield reduction will be the basis for optimum irrigation decision-making. It should also be one which can be monitored quickly and cheaply at numerous points in a field, since the spatial variability of soil moisture and crop conditions may be significant.

Direct measurement of crop stress is a more likely indicator of irrigation requirements than soil moisture measurements because yields are directly related to stress and only indirectly related to moisture conditions (Kramer, 1969). Of course, soil moisture measurements must be made as well. The plant can only signal its need for water while the soil must indicate how much to apply (Teare et al., 1974; Stegman et al., 1976; Stanley et al., 1981).

Soil water tension is not, at present, a suitable indicator. Currently available tensiometers are only effective above about -0.8 bars tension (ASAE, 1980, p. 797). Thus a tensiometer would be operating at the limit of effectiveness when soil moisture is just approaching the range of greatest interest. Development of an inexpensive method of in situ measurement of low matric potentials appears to be a likely research problem in this respect.

Direct measurements of plant-water status (total plant water, matric, turgor and osmotic potential, and stomatal resistance) are accurate indicators of water stress since they integrate the effects of crop species, age, soil moisture, evaporative demand and pest damage. Turner (1981) provides a detailed review of these parameters and their research limitations. A disadvantage of direct measurement of plant-water status is that it is generally expensive due to the sampling frequency, number of samples required, sampling timing, and the time and conditions required to evaluate the samples.

Attempts have been made to correlate plant-water status with more easily measurable variables (radiation, air temperature) with some success (Stegman et al., 1976, Stanley et al., 1981; Smart and Barre, 1973). However, these statistical approaches are generally area specific and require calibration for new crops or plant types.

The calculated ratio of actual evapotranspiration (ET_A) to maximum evapotranspiration (ET_M), termed relative ET, could serve as a quantitative measure of plant stress, since stress will result in reduced ET_A through partial stomatal closure. The departure of ET_A/ET_M from 1.0 occurs at varying soil moisture levels depending upon root density and ET_M levels (Meyer and Green, 1980; Stegman et al., 1976); Mogensen, 1980). Mogensen (1980) demonstrated in a lysimeter study that relative ET began to decrease before leaf water potential in barley. Relative ET was a more sensitive indicator of drought stress than leaf water potential due to the physiological control mechanisms associated with critical leaf water potentials (Denmead and Millar, 1976). Meyer and Green (1980) suggest that there is a reduction in growth prior to the departure of ET_A/ET_M from 1.0. The disadvantage of this indicator is that measurement of relative ET under field conditions is likely to be quite inaccurate.

The use of infrared thermometry presents an indirect method of measuring plant water status. Water stress results in partial stomatal closure and reduction in transpiration, causing canopy temperature to rise above ambient air temperature (Jackson et al., 1977). Thus, the difference between canopy temperature (T_c) and the temperature of the surrounding air (T_a), indicates plant water status. Several researchers have proposed the use of infrared thermometry as a relative measure of plant water stress (Bartholic et al., 1972; Reicosky et al., 1980; Brown, 1974). However, the quantitative relationship between ($T_c - T_a$) and measured plant water stress has not been fully established. A major advantage of canopy temperature as an index of crop stress is that it is adaptable to remote sensing (Idso et al., 1975). If this technique advances to the point that it is possible to reliably monitor conditions in a heterogeneous field at low cost, it could become a valuable tool for optimum scheduling.

2. Dealing with Uncertainty: Filtering

The indicators discussed above must be determined from field measurements if they are to be used for irrigation scheduling. Such measurements are characterized by substantial and largely irreducible uncertainty. The uncertainty can be mitigated by scheduling conservatively, but this tends to reduce expected income. Data filtering is an alternative way of dealing with uncertainty without necessarily reducing expected income. The concept of filtering is briefly outlined below.

Let x(t) represent true values of a state variable x at time t. In the parlance of filtering x(t) is a *signal*. Let $y(t_1)$, $y(t_2)$, ... $y(t_n)$ represent *data*, i.e. measured values of x(t) at times t_1, t_2, ... t_n. These variables are related by

$$y(t_i) = x(t_i) + v(t_i) \tag{1}$$

Where $v(t_i)$ is measurement error (*noise*).

Suppose we wish to use the data to estimate x at some time t*. If $t^* < t_n$ this is a *data smoothing* problem. If $t^* > t_n$ it is a *prediction* problem. If $t^* = t_n$ it is a *filtering* problem (Papoulis, 1965). Thus filtering is concerned with real-time estimation of a state variable using measurements that are noisy.

Under certain circumstances, real-time estimation can be accomplished by a recursive filtering procedure known as Kalman filtering, which can be represented by

$$\hat{x}(t_n) = A(t_n)\ \hat{x}\ (t_{n-1}) + B(t_n)\ y(t_n) \tag{2}$$

where x(t) is a filtered estimate of x(t) and $A(t_n)$ and $B(t_n)$ are weights derived at each time step.

Kalman filtering produces estimates that are optimum in the sense that the expected value of the square of the estimation error is minimized (Papoulis, 1965). When the special circumstances required for Kalman filtering are not present, other filtering algorithms may be devised which will still yield *improved* estimates, though it is often not possible to prove that these estimates are optimal. In general, filtering can be said to yield estimates which are better than the measurements from which they are derived, according to some appropriate criterion.

The improved estimates are arrived at by systematic analysis of the data in conjunction with other information such as models of the physical relationship involved or statistical characteristics of the data. Filters range in sophistication from simple equations, conceived intuitively and developed empirically, to complex and sophisticated algorithms based on the principals of Kalman filtering. Filtering has been used in a variety of applications in which real-time measurements of a state variable are not sufficiently precise. Examples include estimation of the orbital position of satellites, estimation of plankton populations in an aquatic environment (Moore, 1971), real-time determination of atmospheric pollution levels (Soeda and Ishihara, 1974), and a variety of applications in hydrology (AGU, 1978).

Irrigation scheduling has the attributes of a filtering problem. The objective is real-time evaluation of a signal (soil moisture, canopy temperature, etc.). Measurements of these signals tend to be quite noisy due to such things as spatial variability and intrinsic measurement errors. The problem is one of separating the signal from the noise without reducing the clarity of the signal. This effort requires systematic evaluation of the characteristics of both the signal and the noise.

The use of filtering will be illustrated with an example. The example was artificially contrived solely to illustrate the concept and should not be construed as a demonstration of an actual application of filtering for irrigation scheduling.

Relative ET is used as the crop stress indicator in the example because it is particularly sensitive (as noted earlier). On the other hand field measurements of this indicator will be quite noisy, hence filtering might prove useful.

The following relationship between relative ET and soil moisture was assumed

$$\frac{ETA(n)}{ETA(n)} = 1.0 - e^{-c \cdot MR(n)} \tag{3}$$

where ETA(n) = actual daily ET on the nth day
ETM(n) = maximum daily ET on the nth day
MR(n) = relative soil moisture at the beginning of the nth day

$$= \frac{MO - \sum_{i=1}^{m-1} ETA(i)}{MO}$$

MO = available soil moisture at nominal field capacity
C = a constant

This model was derived and calibrated using paired lysimeter data provided by Pruitt (1981). Figure 1 shows this function for various values of C.

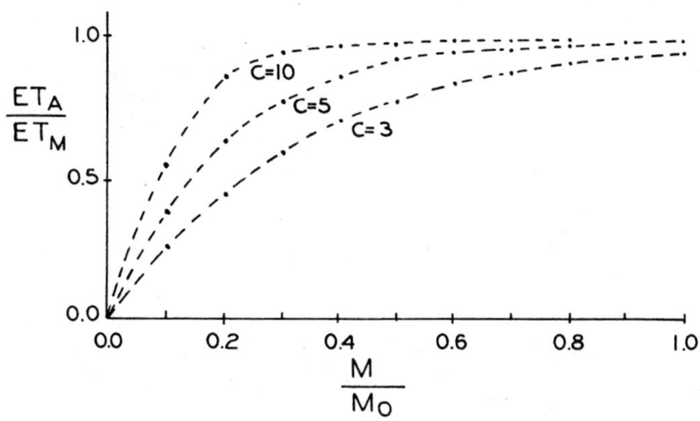

Figure 1. Ratio of actual ET to maximum ET $\left(\frac{ET_a}{ET_m}\right)$

The irrigation decision will be based on cumulative relative ET, defined as

$$CETR(n) = \frac{\sum_{i=1}^{n} ETA(i)}{\sum_{i=1}^{n} ETM(i)} = \frac{D(n)}{\sum_{i=1}^{n} ETM(i)} \tag{4}$$

Where CETR(n) = cumulative relative ET between the last irrigation and the nth day.
D(n) = soil moisture depletion (from a full profile) between the last irrigation and the nth day.

In this example, the decision rule will be to irrigate a field when CETR(n) reaches 0.9.

Measurement of CETR(n) requires measuring D(n) and $\sum_{i=1}^{n} ETM(i)$ separately. It is assumed that D(n) is measured with a neutron probe. Direct measurement of $\sum_{i=1}^{n} ETM(i)$ is impractical under ordinary field conditions, so daily estimates of ETM based on Penman model will be used in lieu of direct measurements. Thus from Equation 4 field measurements of CETR(n) can be calculated by

$$\overline{CETR}(n) = \frac{\overline{D}(n)}{\sum_{i=1}^{m} \overline{ETM}(i)} \quad (5)$$

Where the bar denotes measured values of the variables.

This model is demonstrated using a sequence of hypothetical values of $\overline{CETR}$. These data were synthesized as follows. A 50 day sequence of ETM data from Pruitt's experiment were used. These were assumed to be true values of ETM. Using Equation (3) with c = 5.0 and MO = 7.0 inches a sequence of daily values of ETA were calculated. These were assumed to be true values of ETA. Successive sums of the true values of ETA were computed and these were used as true values of D. Then hypothetical measurements, $\overline{D}(n)$ and $\sum_{i=1}^{n} \overline{ETM}(i)$, were generated by adding simulated random measurement errors to the true values of $\sum_{i=1}^{n}$ ETM and D. Statistical models of these errors were derived from theoretical statistical characteristics of neutron probe measurement errors and observed errors in Penman estimates of evapotranspiration. Finally, hypothetical values of $\overline{CETR}(n)$ were calculated from Equation (5).

These variables are tabulated in Table 1. Figure 2 shows plotted values of $\overline{CETR}$. If these $\overline{CETR}$ data were viewed sequentially, one measurement each day, it would be difficult to detect which day the critical level, 0.9, was reached. The problem then is to use a filter to reduce the uncertainty illustrated by the data in Figure 2.

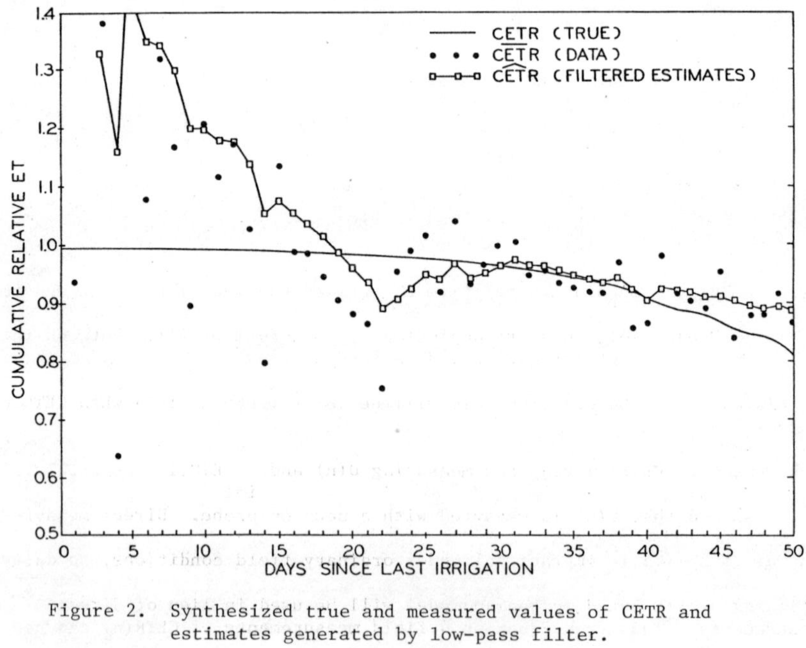

Figure 2. Synthesized true and measured values of CETR and estimates generated by low-pass filter.

Table 1. Synthesized True and Measured Values of Variables in the Filtering Example

	Synthetic True Values					Synthetic Measured Values			Filtered Estimates
Day	ETM (in.)	ETA (in.)	ΣETM (in.)	D (in.)	CETR	$\overline{\Sigma ETM}$ (in.)	$\overline{D}$ (in.)	$\overline{CETR}$	$\hat{CETR}$
1	.07	.070	.070	.070	.993	.105	.098	.934	
2	.13	.129	.200	.199	.993	.209	.517	2.474	
3	.15	.149	.350	.347	.993	.312	.431	1.381	1.335
4	.09	.089	.440	.437	.992	.380	.242	.637	1.160
5	.14	.139	.580	.575	.992	.529	1.207	2.281	1.440
6	.13	.129	.710	.704	.992	.635	.684	1.077	1.350
7	.11	.109	.820	.813	.991	.764	1.008	1.319	1.342
8	.07	.069	.890	.882	.991	.840	.980	1.166	1.298
9	.16	.158	1.050	1.040	.990	1.042	.934	.896	1.197
10	.12	.118	1.170	1.158	.990	1.143	1.382	1.209	1.200
11	.08	.079	1.250	1.237	.990	1.211	1.351	1.116	1.179
12	.17	.167	1.420	1.404	.989	1.396	1.630	1.168	1.176
13	.11	.108	1.530	1.512	.988	1.506	1.541	1.023	1.138
14	.13	.127	1.660	1.640	.988	1.615	1.288	.797	1.053
15	.18	.176	1.840	1.816	.987	1.867	2.111	1.131	1.072
16	.17	.166	2.100	1.982	.986	2.001	1.971	.985	1.051
17	.16	.156	2.170	2.137	.985	2.188	2.152	.984	1.034
18	.20	.194	2.370	2.331	.984	2.454	2.317	.944	1.011
19	.11	.106	2.480	2.437	.983	2.566	2.321	.905	0.985
20	.11	.106	2.590	2.543	.982	2.674	2.348	.878	0.958
21	.19	.182	2.780	2.725	.980	2.879	2.491	.865	0.935
22	.13	.124	2.910	2.849	.979	2.963	2.230	.753	0.889
23	.13	.123	3.040	2.972	.978	3.100	2.948	.951	0.905
24	.20	.189	3.240	3.161	.976	3.289	3.240	.985	0.925
25	.10	.094	3.340	3.254	.974	3.415	3.464	1.014	0.947
26	.11	.102	3.450	3.357	.973	3.529	3.240	.918	0.940
27	.09	.083	3.540	3.440	.972	3.642	3.785	1.039	0.965
28	.17	.157	3.710	3.597	.969	3.862	3.397	.880	0.943
29	.15	.137	3.860	3.734	.967	3.924	3.785	.964	0.949
30	.25	.226	4.110	3.959	.963	4.195	4.173	.995	0.960
31	.22	.195	4.330	4.154	.959	4.407	4.421	1.003	0.971
32	.23	.200	4.560	4.354	.955	4.592	4.336	.944	0.964
33	.21	.178	4.770	4.532	.950	4.852	4.633	.955	0.962
34	.16	.133	4.930	4.665	.946	4.985	4.640	.931	0.954
35	.08	.065	5.010	4.730	.944	5.046	4.666	.925	0.947
36	.18	.144	5.190	4.874	.939	5.169	4.738	.917	0.939
37	.20	.156	5.390	5.030	.933	5.285	4.839	.915	0.933
38	.20	.151	5.590	5.181	.927	5.370	5.198	.968	0.942
39	.32	.233	5.910	5.414	.916	5.796	4.949	.854	0.920
40	.25	.169	6.160	5.584	.906	5.984	5.156	.862	0.905
41	.22	.140	6.380	5.724	.897	6.098	5.967	.979	0.924
42	.22	.132	6.600	5.855	.887	6.358	5.829	.917	0.922
43	.10	.056	6.700	5.911	.882	6.465	5.819	.900	0.917
44	.10	.054	6.800	5.965	.877	6.537	5.789	.886	0.909
45	.24	.125	7.040	6.091	.865	6.813	6.501	.954	0.920
46	.18	.086	7.220	6.177	.855	6.925	5.808	.839	0.900
47	.16	.071	7.380	6.248	.847	7.001	6.147	.878	0.894
48	.10	.042	7.480	6.289	.841	7.063	6.195	.877	0.890
49	.24	.096	7.720	6.385	.827	7.315	6.689	.914	0.896
50	.28	.100	8.000	6.484	.811	7.583	6.583	.868	0.889

Since the signal, CETR, is not stationary this problem is not well suited to Kalman filtering. A filter was therefore developed intuitively. Intially a filter defined by Equation (6) was tried.

$$\hat{CETR}(n) = \alpha \cdot \hat{CETR}(n-1) + \beta \cdot \overline{CETR}(n) \tag{6}$$

in which $\hat{CETR}$ represents a filtered estimate of the signal, and where $0 \leq \alpha \leq 1.0$ and $\beta = 1.0 - \alpha$. If α is greater than β this is referred to as a low-pass filter because a low frequency signal (the slowly changing value of CETR) will pass through the filter with relatively little change while high frequency inputs (the data noise) will be severely attenuated.

After some experimentation, α was set equal to 0.75. The results are tabulated in Table 1 and displayed in Figure 2. The filter eliminated much of the measurement error, though it gave an erroneous indication of the critical level on day 22. Also it did not produce a good estimate during the crucial period between days 40 and 50 when CETR did in fact fall below 0.9.

One problem with this filter is the persistence of prior estimates. In Equation (6) $\hat{CETR}(n-1)$ is, in effect, a preliminary estimate of $\hat{CETR}(n)$, with a weight determined by α. The filter will therefore tend to reproduce prior estimates, and as a result trends in $\hat{CETR}$ will lag behind trends in CETR. If α is reduced the filter will respond more rapidly to changes. But then the interference from noise will be increased and erroneous indications such as occurred on day 22 will be more likely. This illustrates a basic dilemma in filtering, the tradeoff between signal tracking and noise suppression.

Based on our understanding of relative ET we know the signal, CETR, will be monotonically decreasing. But the above model does not take this knowledge into account. To take advantage of this knowledge a filter could be developed which explicitly predicts the decrease in CETR that will take place during the next time step. By predicting the next value of CETR with a model rather than simply using the last estimated value as a prediction, the problem of persistence can be overcome. Such a filter might be defined by

$$\hat{CETR}(n) = \alpha \cdot \hat{\hat{CETR}}(n) + \beta \cdot \overline{CETR}(n) \tag{7}$$

where $\hat{\hat{CETR}}(n)$ is a preliminary estimate of CETR(n) based on the estimate of CETR(n-1) and the current estimate of ETM. This analysis is illustrated by Figure 3. From Equations (3) and (4) it is evident that CETR(n) is a function of $\sum_{i=1}^{n} ETM(i)$. The derivative at time n-1 is indicated by the arrow in Figure 3. A preliminary estimate for time n could therefore be written

$$\hat{\hat{CETR}}(n) = \hat{CETR}(n-1) + \left.\frac{dCETR}{dETM}\right|_{n-1} \cdot ETM(n) \tag{8}$$

(Such a filter was developed by the authors, but space limitations preclude a detailed derivation of the filter and presentation of results.)

This formulation (Equations 7 and 8) illustrates another facet of filtering. The estimates produced by the filter would be generated from a combination of field data and a model of the underlying physical relationships. Thus additional information, in the form of the model, is brought to bear on the problem.

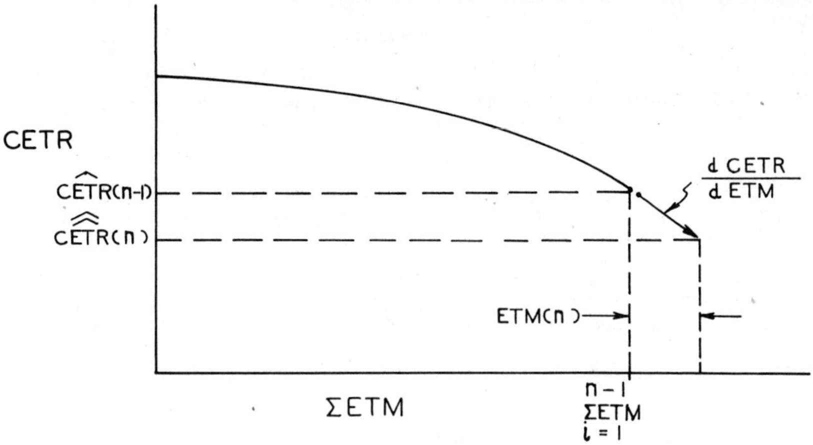

Figure 3. Preliminary Estimation of CETR(n) from $\widehat{CETR}$(n-1) and ETM(n)

It should be noted that the model would probably be biased, since it would have been developed and calibrated under other circumstances. By introducing a biased model the results will be biased as well. If the purpose of the analysis were scientific inquiry it would be best to forgo the additional information rather than bias the results. But the purpose of irrigation scheduling is decision-making, and in that context the additional information may be welcome (English and Orlob, 1978, p. 5).

Summary

It is proposed that the use of crop stress indicators in combination with data filtering techniques will be a basis for successful irrigation scheduling when economic optimization is the goal. A general discussion of the problem of optimum irrigation scheduling has been presented. An optimally irrigated field may be characterized by (a) soil water matric potentials below the level of -0.5 to -1.0 bar at the time of irrigation, (b) more variable soil moisture conditions than found in fully irrigated fields, and (c) more variable crop development. The optimum time to irrigate a crop will be when potential yield has begun to decline. Bearing these facts in mind, it appears that optimal irrigation scheduling will entail the use of crop stress indicators to determine the best timing of irrigations. Ideally such indicators should be sampled quickly and cheaply at several points in a field.

Uncertainty will be an important consideration in scheduling for optimum water use. Filtering techniques might reduce this uncertainty to a manageable level. Two examples of filters were presented to illustrate basic concepts of filtering. The first filter, developed intuitively, demonstrated the process of noise supression and the dilemma of compromising between noise supression and signal tracking. The proposed second filter would be based on underlying physical relationships. Field data would be combined with knowledge of the relationships involved to produce refined real-time estimates of the signal.

REFERENCES

AGU. 1978. "American Geophysical Union Chapman Conference on Applications of Kalman Filtering Theory and Techniques to Hydrology, Hydraulics and Water Resources." University of Pittsburgh.

ASAE. 1980. "Design and Operation of Farm Irrigation Systems," M.E. Jensen, editor. ASAE Monograph No. 3. The American Society of Agricultural Engineers, St. Joseph, Michigan.

Bartholic, J.F., L.N. Namken, C.L. Weigand. 1972. "Aerial Thermal Scanner to Determine Temperatures of Soils and of Crop Canopies Differing in Water Stress," Agronomy Journal 64:603-8.

Brost, C. 1977. U.S. Bureau of Reclamation, Denver, Colorado. Personal communication.

Brown, K.W. 1974. "Calculations of Evapotranspiration From Crop Surface Temperatures," Agricultural Meteorology 14:199-209.

Denmead, O.T. and B.D. Millar. 1976. "Field Studies of the Conductance of Wheat Leaves and Transpiration," Agronomy Journal 68:307-11.

English, M.J. and G.T. Orlob. 1978. "Decision Theory and Irrigation Optimization," California Water Resources Center Contribution No. 174. University of California, Davis, California.

English, M.J., G.L. Horner, G.T. Orlob, J. Erpenbeck, M. Moehlman, R.H. Cuenca and D.J. Dudek. 1980. "A Regional Assessment of the Economic and Environmental Benefits of an Irrigation Scheduling Service," U.S. Environmental Protection Agency, EPA-600/2-80-063. April.

English, M.J. and G.S. Nuss. 1980. "Designing for Deficit Irrigation," Proceedings of the 1980 Specialty Conference, Irrigation and Drainage Today's Challenges, ASCE Irrigation and Drainage Division, Boise, Idaho. July.

Feddes, R.A., P.J. Kowalik and H. Zaradny. 1978. "Simulation of Field Water Use and Crop Yield," Wageningen Center for Agricultural Publishing and Documentation, Wageningen, The Netherlands.

Ferreres, E., B. Amry, J. Faci, A. Kamgar, D.W. Henderson and M. Resende. 1978. "A Closer Look at Deficit High Frequency Irrigation," California Agriculture. August.

Gear, R.D., A.S. Dransfield and M.D. Campbell. 1976. "Effects of Irrigation Scheduling and Coordinated Delivery on Irrigation and Drainage Systems," ASCE National Water Resources and Ocean Engineering Convention, San Diego, California. April 5-8.

Idso, S.B., R.D. Jackson, R.D. Reginato. 1975. "Estimating Evapotranspiration: A Technique Adaptable to Remote Sensing," Science, Vol. 189, pp. 991-2, September 1975.

Jackson, R.D., R.J. Reginato, and S.B. Idso. 1977. "Wheat Canopy Temperatures: A Practical Tool for Evaluating Water Requirements," Water Resources Res. 13:651-56.

Kramer, P.J. 1969. Plant and Soil Water Relationships, McGraw-Hill, San Francisco, California.

Meyer, W.S. and G.C. Green. 1980. "Water Use by Wheat and Plant Indicators of Available Soil Water," Agronomy Journal 72:253-57.

Millar, B.D. and O.T. Denmead. 1976. "Water Relations of Wheat Leaves in the Field," Agronomy Journal 68:303-7.

Miller, D.E. 1977. "Deficit High Frequency Irrigation of Sugarbeets, Wheat and Beans," Proceedings ASCE Specialty Conference, "Water Management for Irrigation and Drainage," Reno, Nevada. July.

Mogensen, V.O. 1980. "Drought Sensitivity at Various Growth Stages of Barley in Relation to Relative Evapotranspiration and Water Stress," Agronomy Journal 72:1033-38.

Moore, S.F. 1971. "The Application of Linear Filter Theory to the Design and Improvement of Measurement Systems for Aquatic Environments," Ph.D. Dissertation, School of Engineering, University of California, Davis, California.

Papoulis, A. 1965. Probability, Random Variables and Stochastic Processes, McGraw-Hill, San Francisco.

Pruitt, W.O. Department of Land, Air and Water Resources, University of California, Davis. May 1981. Personal Communication.

Rawlins, S.L. and Raats. 1975. "Prospects for High Frequency Irrigation," Science, Vol. 188. May.

Reicosky, D.C., D.E. Deaton, and J.E. Parsons. 1980. "Canopy Air Temperatures and Evapotranspiration from Irrigated and Stressed Soybeans," Agricultural Meteorology 21:21-35.

Shearer, M.N. 1978. "Comparative Efficiency of Irrigation Systems," 1978 Annual Technical Conference, The Irrigation Association, Cincinnati, Ohio. February 26-28.

Smart, R.E. and H.D. Bans. 1973. "The Effect of Environment and Irrigation Interval on Leaf Water Potential of Four Horticultural Species," Agricultural Meteorology 12:337-346.

Soeda, T. and H. Ishihara. 1974. "Application of Kalman Filtering Method and n-Observability Principle to the Prediction of the Atmospheric Pollution Levels," Proceedings, Fifth Symposium on Nonlinear Estimation Theory and its Applications, San Diego, California. September 23-24.

Stanley, C.D., T.C. Kaspar and H.M. Taylor. 1981. "Modeling Soybean Leaf-Water Potentials Under Nonlimiting Soil Water Conditions," Agronomy Journal 73:251-54.

Stegman, E.G., L.H. Schiele, and A. Bauer. 1976. "Plant Water Stress Criteria for Irrigation Scheduling," Transactions of the ASAE 19:850-5.

Teare, I.D., E.T. Kanemasu, W.L. Powers and R.L. Stone. 1974. "Physiological Indicators of Water Stress," Evaporation Laboratory, Kansas Agricultural Experiment Station, Manhattan, Kansas. 24 p.

Turner, N.C. 1981. "Experimental Approaches and Techniques for the Measurement of Plant Water Status," Plant and Soil 58:

EVALUATION OF AN IRRIGATION
SCHEDULING SYSTEM FOR VIRGINIA

N. L. POWELL B. B. ROSS F. S. WRIGHT
G. D. JONES V. O. SHANHOLTZ

ABSTRACT

A research program has been initiated in Virginia to develop and test a real-time site-specific water management system for the major cash crops. During the 1980 growing season an irrigation scheduling model for corn was evaluated. A water balance model using soil physical properties, plant characteristics and climatic data was the basis for the irrigation scheduling system. Additional components, such as functions for water use efficiencies and crop susceptibility were added. Climatic data was obtained from an Agro-Environmental Monitoring System which is a remote climatic data collection and transmission system.

Irrigated and non-irrigated field studies were conducted at three sites in the Piedmont and Coastal Plains regions of the state. Soil physical and chemical properties were obtained from each site. Utilizing this, the climatic data and the crop susceptibility function, the system was updated daily and recommendations were made concerning the advisability of irrigation. Gravimetric soil moisture samples were taken during the growing season for the purpose of model verification and evaluation. Treatments in the irrigated and non-irrigated field plots included population density, conventional tillage, minimum tillage and under-row ripping.

Increased yields resulted from irrigation at all sites as the 1980 growing season was one of the driest on record. The amount of irrigation water applied ranged from 295 mm at one site to 350 mm at another. Irrigated yield increases ranged from 4.9 to 8.5 Mg/hectare over non-irrigated yields. At one site irrigated yields were over five times that of non-irrigated yields.

The second test year of the irrigation scheduling system will be conducted in 1981 with field studies at an additional one or two sites within the state.

The authors are: N. L. Powell, Assoc. Prof., Dept. of Agronomy. VPI&SU; B. B. Ross, Res. Assoc., Dept. of Ag. Engr., VPI&SU; F. S. Wright, Ag. Engr., Tidewater Research and CEC; G. D. Jones, Director, Piedmont Research Station; and V. O. Shanholtz, Assoc. Prof., Dept. of Ag. Engr., VPI&SU, Blacksburg, VA.

A Research Tool for Evapotranspiration Measurements for Model Validation and Irrigation Scheduling[1]

D. C. Reicosky[2]

INTRODUCTION

Proper evaluation of the water resources and more scientific management of the water requires the development of tools that will provide satisfactory measures of crop water use. Tools frequently used in evaluation of evapotranspiration (ET) are the weighing lysimeter and the assimilation chamber techniques reviewed by Tanner (1967).

The lysimeter is the only hydrologic method in which the experimenter can obtain accurate and continuous measurements of ET. The lysimeter has been used for short-term measurements and can provide great accuracy under certain conditions. However, the expense of the lysimeter as well as the large uniform crop area required limit its usefulness when trying to evaluate water use for different soil and water management practices. Another method of measuring ET on field plots utilizes the portable chamber technique. Examples of such chambers are described by Musgrave and Moss (1961), and Puckridge (1978). Peters et al. (1974) described an automated travel system that measured photosynthesis and ET simultaneously. Reicosky and Peters (1977) have described a portable chamber for the rapid measurement of ET on field plots. While there has been some question about the validity of chamber measurements based on the discussion of Businger (1963), e.g. convection and temperature effects, the relative importance of these errors in ET measured with a portable chamber has not been fully evaluated. The objective of this work is to provide an overview of recent ET data collected with a portable chamber for various soil and water management practices.

MATERIALS AND METHODS

Chamber Operation

Construction details of a portable chamber that permits measurements over 76- and 102-cm row spacings for small corn and soybeans have been previously discussed by Reicosky and Peters (1977). A slightly modified version of the portable chamber currently being used in Minnesota is shown in Fig. 1. The chamber contains four fans for complete mixing of the air and is front mounted on a vertical lifting mechanism attached to the front end loader of a medium sized farm tractor. The tractor transports the chamber and associated recording equipment from plot to plot along with a portable generator that enables measurements in remote field locations. ET measurement areas on the field plots were premarked with reference stakes to enable repeated measurements and the soil around the perimeter is leveled to enable the chamber to form a seal with the soil surface. The tractor with the portable chamber in the up position is maneuvered until the chamber reference points align with reference stakes adjacent to the measurement area and then the tractor brake is set. With the chamber fans running continuously, the strip chart recorder is started to record the air temperature (T_a) and the wet bulb temperature

---

1/Contribution from the North Central Soil Conservation Research Laboratory, Agricultural Research Service, USDA, Morris, MN, in cooperation with the Minnesota Agricultural Experiment Station, Sci. Jour. Series No. 11,819.

2/Soil Scientist, ARS, USDA, Morris, MN.

Fig. 1. Photograph of the portable chamber currently being used in Minnesota.

is started to record the air temperature (T_a) and the wet bulb temperature (T_w) of the aspirated psychrometer. Once the initial dry and wet bulb temperatures are constant (about 20 sec.), the chamber is then lowered over the measurement area (requires about 5 to 8 sec. depending on initial height). The strip chart is then marked to indicate the time of chamber-soil contact and the wet and dry bulb temperatures are recorded for an additional 60 to 80 sec. The chamber is then raised and the recorder stopped in preparation for backing the tractor and chamber off the measurement area and moving to the next area. For each ET measurement, the time that the chamber makes contact with the soil, plot identification, cloud cover, solar radiation, and other pertinent notes are recorded. Once the plot reference stakes have been installed, the chamber requires only one operator.

Vapor Density Measurement

The dry and wet bulb temperatures at the time the chamber hits the soil surface and exactly 60 sec. later are used as the initial and final temperatures in the calculation of corresponding vapor densities. These temperatures are used to calculate the ambient vapor pressure using the psychrometric equation and the saturation vapor pressure at T_w using the Richards' (1971) equation. The initial and final ambient vapor pressures are substituted in the ideal gas equation with the appropriate constants to yield initial and final vapor density expressed in g/m^3. Knowing the initial and final vapor density, the time interval, the chamber volume and the soil area, the ET is then calculated. Further details on the calculation of the ET are given by Reicosky and Peters (1977).

Recent work in Minnesota comparing a modified chamber with a weighing lysimeter indicates that the measurement time interval should probably be decreased from 60 sec. to 30 sec. (Reicosky et al. 1981). As the plants transpire, there is a slight build up in the water vapor density within the chamber and the wet bulb temperature does not continue to increase in a linear fashion but tends to become asymptotic as saturation is approached. Thus, it is essential that the T_a and T_w measurements be made as rapidly as possible before the T_w changes become nonlinear.

REVIEW OF PREVIOUS WORK

Samples of Data Collected with the Portable Chamber

Initial calibrations of a portable ET chamber and the techniques used were described in detail by Reicosky and Peters (1977). The portable chamber was used to measure rate of water loss from plants grown in a hydroponic system designed to accurately measure plant water uptake. The results showed a linear relationship between the transpiration measured by the chamber and the absorption measured by the hydroponic system indicating that the portable chamber was indeed measuring transpiration. In this calibration work the soil surface was covered with polyethylene to minimize soil evaporation. Normally the plastic is not used on field plots so the measurement includes soil evaporation and transpiration (ET). In this calibration study, the variation in transpiration was apparently the result of not reading temperatures accurately from a strip-chart recorder. A brief error analysis indicated a possible overall error of 19 percent and a probable error of 11 percent if temperatures were read to the nearest 0.1°C.

The effect of a drought on ET of field grown soybeans was discussed by Reicosky and Deaton (1979). Soybeans were grown on a Norfolk sandy loam to evaluate the effects of irrigation on yield and to determine the combined effects of soil water potential and evaporative demand on ET of irrigated and nonirrigated soybeans. While only small differences in midday leaf water potential were found between the cultivars, ET on the nonirrigated plots after 25 and 32 days without significant rainfall ranged from 40 to 60 percent of the ET on the irrigated plots. The soil water potential profiles indicated significant water extraction to a depth of 153 cm on the nonirrigated plots near the end of the drought. The small differences in the minimum leaf water potential indicated that the plants had partially adjusted to the drought. However, there was partial stomatal closure as evidenced by visible wilt symptoms and lower ET values measured with the portable chamber.

The relationship between canopy air temperatures and ET from irrigated and stressed soybeans was discussed by Reicosky et al. (1980). As soil water stress became more severe, canopy air temperatures within nonirrigated soybeans increased above those within an irrigated canopy. When plant wilt symptoms indicated severe stress, ET on nonirrigated plots decreased to 40 to 70 percent of that on the irrigated plots. Within canopy air temperatures increased 4 to 6°C. The increase in canopy air temperature and the decrease in ET were related to stem diameter shrinkage.

Evapotranspiration measurements were made on selected days in the soybean growing season as part of a study on the effect of row spacing and soil water supply on water use in Western Iowa (Mason et al. 1980). In this work the transparent material used on the chamber was 3.18 mm thick plexiglass. Plexiglass reduced the light intensity inside the chamber 8 percent and may have reduced the actual ET. However, the effect should be similar across all treatments. An example of ET and solar radiation data is presented in Fig. 2 for the two different row spacings on 11 July 1979 (58 days after planting). The last significant rainfall was 18 mm 21 days earlier and the last irrigation was 5 cm applied 9 days earlier. Because of the scatter, the ET data were smoothed to show the diurnal trend using a 1-2-3-2-1 weighted running average as represented by the solid lines. Much of the scatter in the ET data was related to variations in solar radiation resulting from intermittent clouds at the time of the measurement. For the 25-cm row spacing, the midday ET was 0.5 mm/hr on the irrigated plots, while that on the nonirrigated plots was about 0.4 mm/hr. For the 100-cm spacing there was similar scatter in the data with less difference between the irrigated and nonirrigated treatments. Daily ET estimates were obtained from the area under the weighted running average line between sunrise and sunset and showed 4.9 and 3.4 mm/day for the 25-cm row spacing irrigated and nonirrigated treatments,

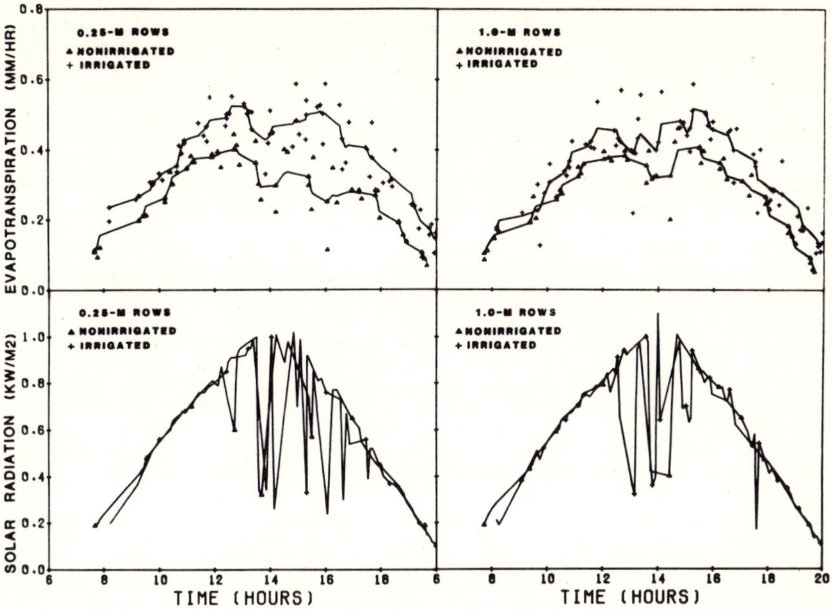

Fig. 2. Summary of evapotranspiration and solar radiation versus time (CDT) for irrigated and nonirrigated soybeans in 0.25 and 1.0 m row spacings on 11 July 1979. (From Mason et al. 1980.)

respectively. For the 100-cm row spacing the corresponding daily ET was 4.4 and 3.5 mm/day. As can be seen from this data, the largest differences in ET were due to irrigation on the 25-cm row spacing.

The small difference in daily ET between the 25-cm and the 100-cm irrigated treatments was apparently due to the larger leaf area index (LAI) on the 25-cm spacing because the soil water data indicated that the soil water status was essentially the same (Mason et al. 1980). The LAI on the 25-cm row spacing was 3.4 and 2.5 for the irrigated and the nonirrigated treatments, respectively. For the 100-cm row spacing, the irrigated and nonirrigated treatments had a LAI of 2.7 and 2.5, respectively. The small differences in ET measured with the chamber were apparently related to the differences in LAI. When viewed from above, the 100-cm row spacings showed considerable soil surface exposed with none apparent on the 25-cm row spacing. The small difference between the irrigated 100-and 25-cm row spacings treatments suggests the differences in ET were primarily the result of differences in leaf area index. For the nonirrigated treatments, the 25-cm row spacing had slightly lower ET in the afternoon, apparently as a result of a slightly higher water use rate earlier in the season, thus depleting soil moisture sufficiently to lower ET more than the 100-cm nonirrigated treatment (Mason et al. 1980). The narrower row spacing appeared to use more water earlier because of the earlier soil coverage (larger LAI) which resulted in an earlier decrease in ET on the nonirrigated treatment.

CURRENT RESULTS

A recent study was conducted to compare the portable chamber directly with a weighing lysimeter at the Minnesota Agricultural Experiment Station (Reicosky et al. 1981). Measurements were made when the alfalfa was approximately 70 cm tall. The lysimeter and surrounding area were irrigated with 5 cm of water the preceding day. Ground cover was complete and full canopy had developed with no visible differences between the canopy stature in the lysimeter and the measurement area for the portable chamber which was within 10 m of the weighing lysimeter. Measurements were made with the portable chamber and the output of the lysimeter was printed on 10-minute intervals.

The diurnal patterns of ET from the lysimeter and the portable chamber were compared on a relatively clear day by taking the hourly average using the 1-2-3-2-1 weighted running average of the 10-min readings. Short term output from the weighing lysimeter readout device was erratic due to electrical noise and required additional smoothing of hourly data. ET measured by the weighing lysimeter and the portable chamber went through the expected diurnal patterns when compared on an hourly basis. The diurnal ET pattern was related to the solar radiation and the vapor pressure deficit. A maximum ET of 0.85 mm/hr for the lysimeter and 0.80 mm/hr for the portable chamber was obtained. The relationship between the hourly ET measured by the lysimeter and the chamber is shown in Fig. 3. The data indicate fair agreement. Daily ET values estimated between 0500 and 2100 CST from the area under the curves of

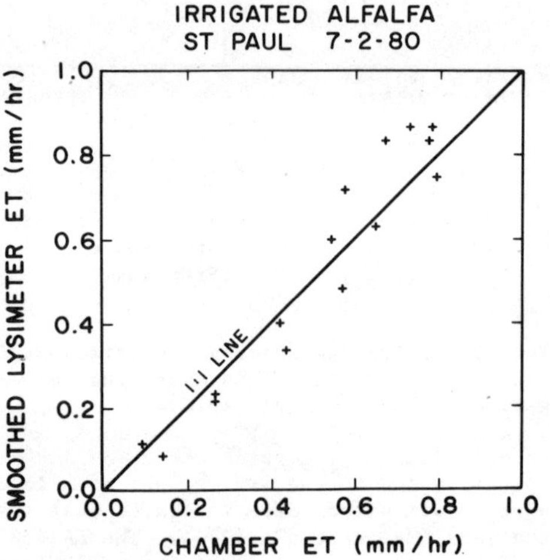

Fig. 3. Hourly averages of evapotranspiration from irrigated alfalfa measured by the weighing lysimeter versus that measured by the portable chamber on 2 July 1980 (from Reicosky et al. 1981).

the weighted running average line for ET versus time were 8.0 mm/day for the portable chamber and 7.8 mm/day for the lysimeter. These results are in agreement with those of Puckridge (1978) who concluded the enclosure method is reasonable for comparative measurements of water use on different treatments. In the work with irrigated alfalfa, both hourly and daily values of the ET showed reasonable agreement and indicated that the portable chamber approximates the ET being measured by the weighing lysimeter.

DISCUSSION

Limitations of the Portable Chamber

The environment inside the portable chamber is often not considered representative of the natural environment of the plants since the radiation exchange and the turbulent transfer within the chamber are altered as discussed by Businger (1963). The ET obtained by the portable chamber is a point measurement and integrated values on a daily basis or a long term basis require repeated readings and hence is laborious. The frequency of measurements to obtain daily water use depends on the type of day. On a relatively clear day, one measurement per hour would be adequate to characterize the daytime ET. However, with intermittent clouds passing over and rapidly changing vapor pressure more frequent measurements are required. The portable chamber is not practical for anything more than daily values (daytime only). Because the measurements must be made during the daylight hours, there may be errors involved in trying to estimate total water use from point measurements. These errors may be propogated through the season and result in substantial differences in seasonal water use. In humid climates, the presence of dew in the early morning hours make the measurements and interpretation of ET difficult.

Uses of Evapotranspiration Data Obtained with a Portable Chamber

While the portable chamber has the limitations discussed above, results obtained so far indicate general agreement in measuring plant water use from a hydroponic system and weighing lysimeter. ET measured in row spacing and irrigation studies was in general agreement with expected results based on theoretical considerations. While the portable chamber may not have the precision of the weighing lysimeter, it does have the portability for measuring many different treatments where the number of weighing lsyimeter would be impractical. The combination of the weighing lysimeter and the portable chamber should allow for the maximum utility of water use data. The portable chamber can be used to measure short term ET on several different treatments while the weighing lysimeter can be used to measure long term ET on a selected treatment. These two methods combined with the appropriate modeling technique to translate the short term chamber ET data to more meaningful long term ET data should result in more useful information.

The portable chamber itself is not practical for determining when to irrigate on a commercial farm but can serve as a useful research tool in the development of irrigation scheduling criteria. Irrigation scheduling criteria are interdependent on plant properties, soil properties and on the meteorological conditions. Any one of these scheduling criteria could be selected for determining when to irrigate. For example, one could select the air temperature - foliage temperature difference as the criteria. The portable chamber can measure ET on a treatment and on a check plot (well-watered control) to determine when the treatment ET has just decreased below the check plot. Corresponding measurements of leaf temperature and air temperature then enable one to determine the critical value for this difference and thus can be used for determining when to irrigate. Further, the associated soil water and plant data at this critical ET can then be established for maximum utilization of the applied water. The decrease in ET on plots with soil water stress then can be related to soil water potential and plant rooting characteristics and provide useful information on the critical levels of soil and plant parameters for irrigation scheduling.

The portable chamber can be useful in validation of the ET portion of hydrologic models on a watershed scale. Many of the hydrologic models work on a scale that would require numerous weighing lysimeters at various locations. As long as the measurement area and crop canopy is representative of the watershed that is being evaluated, ET can be measured with the portable chamber at remote locations and provide point data necessary to validate the ET models on a short term basis. As long as plot locations are in close proximity, ET measurements can be made at different places in the watershed to determine ET variations based on soil and plant characteristics. While the chamber can be useful in validating ET models in the short term, ET measurements for periods of several days would be laborious and time consuming.

The reasonable agreement between the portable chamber and lysimeter data suggest that the portable chamber can be a useful tool in water management studies. The portability of the chamber, the accuracy and the repeatability of the measurements show utility in evaluating ET models. Further improvements in sensors and in the automation of the data collection should result in improved accuracy. Further study using the portable chamber should lead to better irrigation scheduling criteria for a wide range of soil and plant management systems.

REFERENCES

1. Businger, J. A. 1963. The glass house (greenhouse) climate. In W. R. van Wijk (ed.) Physics of plant environment. North Holland Publishing Co., Amsterdam. pp. 277-318.

2. Mason, W. K., H. M. Taylor, A. T. P. Bennie, H. R. Rowse, D. C. Reicosky, Y. Jung, A. A. Righes, R. L. Yang, T. C. Kaspar, and J. A. Stone. 1980. Soybean row spacing and soil water supply: Their effect on growth, development, water relations, and mineral uptake. Advances in Agriculture Technology. USDA-SEA-NC-5, pp. 1-59.

3. Musgrave, R. B., and D. N. Moss. 1961. Photosynthesis under field conditions. I. A portable closed system for determining net assimilation and respiration of corn. Crop Sci. 1:37-41.

4. Peters, D. B., B. F. Clough, R. A. Garves, and G. R. Stahl. 1974. Measurement of dark respiration and photosynthesis in field plots. Agron. J. 66:460-462.

5. Puckridge, D. W. 1978. A comparison of evapotranspiration measurements of crop communities using lysimeters and assimilation chambers. Aust. J. Soil Res. 16:229-236.

6. Reicosky, D. C., and D. E. Deaton. 1979. Soybean water extraction leaf water potential, and evapotranspiration during drought. Agron J. 71:45-50.

7. Reicosky, D. C., D. E. Deaton, and J. E. Parsons. 1980. Canopy air temperatures and evapotranspiration from irrigated and stressed soybeans. Agric. Meteor. 21:21-35.

8. Reicosky, D. C., and D. B. Peters. 1977. A portable chamber for rapid evapotranspiration measurements on field plots. Agron. J. 69:729-732.

9. Reicosky, D. C., B. S. Sharratt, J. E. Ljungkull and D. G. Baker. 1981. Comparison of alfalfa evapotranspiration measured by a weighing lysimeter and a portable chamber. Agron. J. (In press).

10. Richards, J. M. 1971. Simple expression for the saturation vapor pressure of water in the range of -50 to 140°C. The Brit. J. Appl. Phy. 4:L15-L18.

11. Tanner, C. B. 1967. Measurement of evapotranspiration and irrigation of agricultural lands. Edited by Hagen, Haise and Edminister. Amer. Soc. of Agron. Monograph No. 11, pp. 534-574.

IRRIGATION SCHEDULING WITH SOIL INSTRUMENTS: ERROR LEVELS AND MICROPROCESSING DESIGN CRITERIA

J. W. Cary 1/

Two criteria for deciding when a crop should be irrigated are: (1) the depletion of water in the root zone to some predetermined amount, or (2) the decrease of water potential at some given depth to a predetermined level. The value one chooses for either of these criteria to indicate that irrigation is needed will depend on soil properties, crop rooting characteristics and stage of plant growth. Functional relations between these two criteria and production are not yet known quantitatively, thus one cannot say that either approach is inherently better than the other. The effective application of either requires experience and judgment.

Recent years have seen significant progress in scheduling irrigation using meteorological data to calculate the depletion of water in the root zone. The daily potential evaporation from a full cover reference crop can be calculated within a few percent using measurements of air temperature, humidity, solar radiation and wind run. Given an appropriate crop coefficient curve, the evapotranspiration can also be estimate and the soil water depletion known with varying degrees of accuracy. As an alternative, the rate of soil water depletion may be directly measured with a neutron meter or by gravimetric soil sampling. Gear et al. (1977) used a neutron meter to measure soil water on successive dates and projected soil water depletion with a straight line to a level where replenishment would be needed. This gave estimates of the number of days until irrigation.

Tensiometers, resistance blocks, thermoconductivity sensors, psychrometers and related instruments have been occasionally used or proposed for use in automatically starting irrigation at some given water potential. Tensiometers and gypsum resistance blocks have been available for many years to help decide when the soil should be irrigated. Fischback (1978) reported results of scheduling the irrigation of corn by several different methods including resistance blocks and a meteorological approach. He tended to favor the blocks.

At the present time, a farm manager may schedule irrigation with soil water potential instruments. Based on experience, he will extrapolate the soil water change expected in the next few days, and arrive at a projected date for irrigation. The recent evolution of microprocessors suggests that a system might be designed that would automatically read soil water potential instruments and predict the day to irrigate using an appropriate algorithm. This could lead to a level of sophistication for predicting irrigation frequencies comparable to that developing for computer scheduling with microclimate data (Wright and Jensen, 1978).

The general patterns of change in soil water potential with time at a given depth are shown in Fig. 1.

1/ Soil Scientist, Agriculture Research Service, U. S. Department of Agriculture, Snake River Conservation Research Center, Kimberly, Idaho 83341.

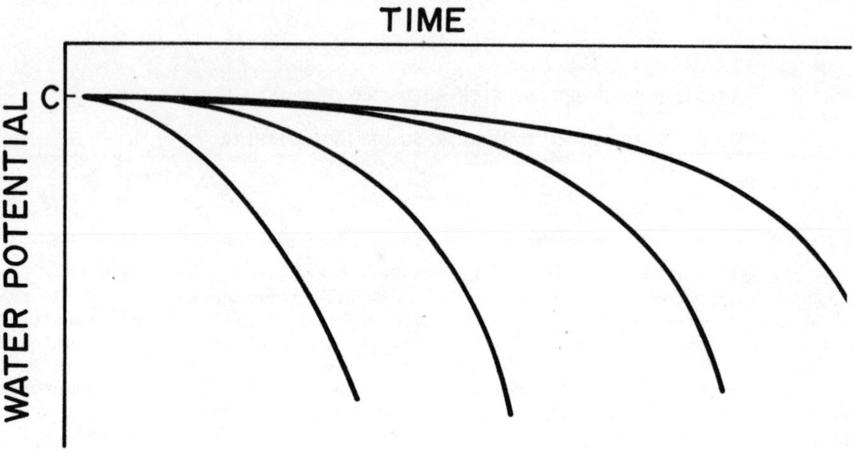

Fig. 1. A Family of Curves Illustrating the Effects of Various Evapotranspiration Rates on the Relation Between Time and Soil Water Matrix Potential, τ, at a Given Depth.

This family of curves can be approximated by the empirical function

$$\tau = At^n + C \tag{1}$$

where t is time in days and τ is soil matrix potential in kPa. The constant n depends mostly on soil pore size distribution and the type of sensor, while the constant A is affected more by the rate of soil water depletion and C is the intercept at $t = 0$, i.e., immediately following irrigation. Note that Eq. 1 is a straight line using variables τ and (t^n). Consequently a value for n can be found for individual soils by measuring water potential changes during periods of evapotranspiration and choosing an n that gives the most consistent linear plots. Since the constants in Eq. 1 can be calculated from appropriate data, a microprocessor could be programmed to project irrigation by extrapolating time to some predetermined water potential. The feasibility of such an undertaking requires an analysis of the inherent errors with respect to the errors encountered in scheduling with microclimate data and with respect to the level of error that is practical for the grower. Specifically, it comes down to questions on the applicability of Eq. 1, the accuracy and reliability of soil water potential sensors, and the spatial variability of soil water in the field.

EXPERIMENTAL PROCEDURES

Data were collected during two growing seasons on four plots of Portneuf silt loam soil described in detail by Cary and Rasmussen (1979). Each plot was 200 meters long and 9 meters wide. Corn, beans, sugarbeets and grass were grown the first year; beans and sugarbeets followed the second year. The plots were irrigated with corrugates (except the grass which was border flooded). Each plot was sampled and instrumented on both ends such that data collection sites were about 170 meters apart.

There were 8 data collection sites the first season. Each had two permanent tensiometers at the 30-cm depth. There were commercial units with 45-cm long plastic cylinders connecting the ceramic cups to vacuum gauges. Three gypsum resistance blocks2/ were also installed in the rows at the 30-cm depth within a 2-m radius. The resistance of each block was measured five days a week with a 1K Hertz electrical conductivity bridge. Soil temperatures around each set of blocks were also measured. Gravimetric soil water measurements at depths of 15, 30, and 45 cm were made from two cores taken about 2 m apart twice weekly at random locations in the rows near the blocks and tensiometers.

In the second season, the blocks were placed at both 15- and 30-cm depths and their resistances were measured twice a week. Three portable rapid response tensiometers3/ were also inserted twice weekly to measure water potential at the 30-cm depth. These tensiometers were placed at random not farther than 2 m apart in crop rows near the blocks.

Care was taken to irrigate the plots as uniformly as possible. Fertility and cultural practices were in accord with local recommendations and practices.

At the end of the growing season, 4 sites, 2 at each end of the field, were sampled, taking four undisturbed cores from each site. Slices were taken from these cores at the 26- to 34-cm depths and individual moisture desorption curves measured for each core. These data were used to calculate pore size distribution indexes by the method of Cary and Hayden (1974).

RESULTS

The gravimetric soil water measurements were used to assess the spatial variability within the plots and to compare the variability shown by the tensiometers and blocks. Methods of characterizing soil spatial variability are not yet very well developed, though this is being addressed by a number of soil scientists (Rao et al., 1979 and Western Regional Research Committee 155). In this case, the standard deviation was calculated for two or more observations that should have been identical. This value was divided by the mean of the observations to get the coefficient of variation. The average of all the coefficients of variation was then used to characterize the variability (Table 1). This approach reduces the dependence of standard deviation on the range of the data observations since the standard deviation of water potential increases rapidly as the potential becomes more negative.

The standard deviation of the water content increases as the water content increases, but as pointed out by Ben-Asher (1979), standard deviation in general for uniform soils is about 10% of the water content. Nielson et al. (1973) found average standard deviation of volumetric water contents between 5 and 7% in a field study, indicating a coefficient of variation range of 13 to 20%. Cassel and Nelsen (1979) also reported the coefficients of variation of volumetric water contents ranged from 8 to 25% in an intensive field study.

Entries 1, 3, 7 and 9 in Table 1 are measures of the short distance spatial variability of soil in the test strips. Average values of water potential from each of these local observation sites were averaged and their means used to characterize the overall spatial variability of the study area, i.e., entries 2, 4, 5, 8 and 10. Averaging several localized measurements to get means for characterizing the overall spatial variability reduces the error caused by the inherent variability of the measuring instrument as demonstrated

---

2/ Beckman Instrument Company, Cedar Grove, New Jersey.

3/ Soil moisture probe Model 2900, Soil Moisture Equipment Corporation, Santa Barbara, California.

Trade names and company names are included for the benefit of the reader and do not imply any endorsement or preferential treatment of the product by the U. S. Department of Agriculture.

Table 1. Coefficients of Variation Associated with Measured Values of Soil Water Potential, Water Contents on a Weight Basis, and Soil Pore Size Distribution Indexes.

Sample	Mean Coefficients of variation	Number of observations	Range of samples
1. Duplicate water contents from cores less than 2 m apart.	4.7%	1144	9 to 30% by wt.
2. Mean water contents from 8 separate field sites.	5.4%	582	10 to 28% by wt.
3. Pore size index from undisturbed samples at 30 cm within a 2-m diameter.	10.5%	16	2.4 to 3.6
4. Pore size index means representing 4 widely separated field sites.	5.9%	4	3.0 to 3.4
5. Water potentials from mean water contents at 30 cm representing 8 field sites.	20.6%	186	-20 to - 72 kPa
6. Calibration data of 28 individual gypsum blocks.	12.5%	136	-30 to -150 kPa
7. Individual blocks making up the means shown in Table 2, columns 2 and 3.	25.7%	306	-20 to -900 kPa
8. Data from columns 2 and 3 in Table 2.	16.6%	102	-30 to -800 kPa
9. Rapid response tensiometers separated by leas than 2 m.	11.9%	384	- 5 to - 76 kPa
10. Tensiometer means, representing 4 widely separated field sites.	13.3%	106	- 1 to - 70 kPa

by the reduction in the coefficient from entries 7 to 8. The coefficient in entry 7 came from 12 blocks, 3 each at 4 widely separated sites in the study area. The coefficient in entry 8 was based on the means of 3 localized blocks at each of the 4 separated sites.

Water potential was estimated from the gravimetric water content at the 30-cm depth, assuming a single water desorption function, Fig. 2, for all four plots (entry 5, Table 1). The coefficient of variation of these potentials was 20.6% which was greater than the coefficient of variations of the potentials measured with blocks or tensiometers (entries 8 and 10). Consequently, it appears the inherent inaccuracies in measuring soil matric potential with gypsum resistance blocks or tensiometers may be no greater than the inherent spatial variability in the field. Particularly with blocks, the average resistance of several placed near one another may be used to reduce the effect of their inherent variability.

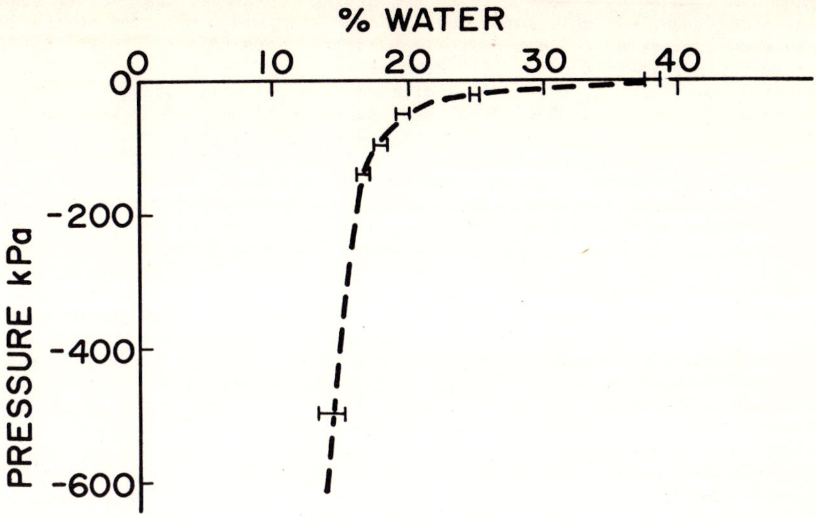

Fig. 2. Water Release Curve for the Soil Studied, Percent Water Dry Weight Basis as a Function of Pressure in the Desorption Chamber. Brackets Show the Spread of Mean Water Contents from the Four Sampling Sites. The Pore Size Distribution Indexes Associated With These Data are Given in Table 1.

DISCUSSION

Several problems associated with automating the tensiometers and blocks with a microprocessor to read out projected irrigation dates were noted during the experiment and analysis of data. The permanent plastic tube tensiometers were unsatisfactory for automation because they required weekly service and were sluggish when the soil was drier than -60 kPa. They had to be placed 30 cm deep to remain operative for at least the first two-thirds of some irrigation cycles. The rapid response tensiometers were better, but even they had to be recharged once or twice during the season. Their effective range was only down to a bit less than -70 kPa (elevation 1,130 m). Their mobility and rapid response time were advantages insofar as characterizing conditions in the field, but considerable care was required to install them, particularly when the soil surface was dry and slaked into the access holes.

The gypsum resistance blocks also have several inherent problems. They are temperature dependent. The empirical equation

$$R_{22} = [(T^{0.011} - 1)(T-22) + 1] R \qquad (2)$$

was found from measurements made in a controlled temperature room. It was used to correct the observed resistance to the 22°C calibration temperature. A second empirical equation was then developed for the soil water potential

$$\tau = 0.0217 R_{22} + 69.84 \ln R_{22} - 3.245 \sqrt{R_{22}} - 310.5 \qquad (3)$$

where R is the measured resistance in ohms, T is the temperature of the block °C and R_{22} is resistance corrected to 22°C.

Gypsum blocks also have some wetting and drying hysteresis that may be significant under transient soil water conditions. Problems were noted at the 30-cm depth when irrigation did not quite increase the potential to -30 kPa before rapidly falling again due to soil water extraction. In essence, the blocks did not always rewet to the level indicated by gravimetric samples. This problem was less at the 15-cm depth because the soil water content rises higher following rain and light irrigations allowing a more complete block response. As a consequence, blocks at the 15-cm depth passed through a wider range of water potentials during each drying cycle than blocks at the 30-cm depth or tensiometers at any depth. This is an advantage for automation and data reduction with a microprocessor.

The mean water potentials from the three temperature corrected blocks at the 15-cm depth are shown in Table 2 for the upper and lower ends of the beet and bean plots during the second season. The predicted lengths of the drying cycles are also shown at various times during each cycle for the lower and upper ends (columns 4 and 5). The predictions are from Eq. 1 taking $n = 3$, C (i.e., field capacity) as -30 kPa and using the average water potential from the three blocks on day, t, to find the value of A. With exception of the first few days following irrigation, this method predicted the length of drying cycles within one or two days of those observed, even though the cycles ranged from five to 25 days. This method requires only a portable nonpolarizing meter to measure the resistance of the blocks and a simple hand held programmable calculator. The manager must enter the resistance, number of days since irrigation, an estimate of the soil temperature at 15 cm, the field capacity, C, and the potential at which he wishes to irrigate. He will then receive the projected number of days until irrigation. This simple predictive method using only the current day block reading may encounter problems following a light rain that lowers the resistance but does not bring the blocks all the way back to field capacity. In this case some judgment will be needed by the operator concerning the appropriate value of A and t. With respect to the data in Table 2, the only significant rainfall during the growing season was 0.7 cm on May 23, 0.9 cm on June 18, and 1.9 cm on August 13-15. Irrigation can also be scheduled with a programmable hand held calculator using weather data (Kanemasu et al., 1978). However, daily weather records are needed for input as well as specific information on crop and soil conditions.

The last column in Table 2 gives the drying cycle length calculated from a linear regression of Eq. 1 using τ and (t^3) as variables. This method gives values for both A and C. Input data for each day was the mean potential of all six resistance blocks in the irrigation strip, in this case not corrected for temperature. Regression analysis was started after each irrigation. For fewer than five days, the length of the cycle was calculated from Eq. 1 taking time equal to one day, the current mean block resistance as C, and using the slope A from the previous cycle. Again, after the first few days, the regression method gave cycle lengths in good agreement with the observed values; i.e., the calculated values were generally within the limits of uncertainty due to spatial variation between the ends of the plot. Neglecting the temperature correction made no significant difference until September when the predicted cycle became several days too short. Soil temperatures had fallen to 12-14°C compared to 20-22°C throughout most of the summer. Possibly automation and storage of the daily block resistances for use in regression analysis would reduce the prediction error during the first few days following irrigation. In any case, using the regression analysis in a microprocessor that receives daily data input eliminates the need for operator judgment following a rain or light irrigation where the blocks do not go all the way back to a field capacity reading. The processor would treat any significant drop in resistance as the start of a new irrigation cycle and use the slope A from the previous cycle to project irrigation dates until a few days pass and provide a more current data base for regression.

Table 2. Predicted and Observed Lengths of Time Required for the Soil at 15 cm to Reach Various Water Potentials Following Irrigations for the Lower and Upper Ends of the Field.

Irrigation date and crop	Days after irrig.	Observed Potential, kPa		Predicted Number of days until the specified H_2O potential was reached		
		Lower end	Upper end	Lower end	Upper end	Combined
22 May	7	- 33	- 33	22	22	-
Beets	10	- 46	- 46	18	18	19
	14	- 83	- 83	17	17	18
	17	-110	- 91	18	19	18
Observed number of days for the H_2O potential to fall to the specified level of -120 kPa →				18	19	
11 June	1	- 29	- 35	-	4	28
Beets	4	- 37	- 60	14	8	10
	8	- 49	- 71	19	15	17
	11	- 63	- 84	22	19	21
	15	-118	-167	22	19	21
	18	-205	-308	21	18	20
	21	-389	-470	19	18	18
Observed number of days for the H_2O potential to fall to the specified level of -300 kPa →				20	18	
03 July	2	- 37	- 54	8	5	22
Beets	6	- 87	-112	11	10	11
	10	-320	-310	11	11	11
	14	-799	-557	11	12	12
Observed number of days for the H_2O potential to fall to the specified level of -400 kPa →				11	12	
18 July	2	- 41	- 57	6	5	14
Beets	7	- 79	- 94	14	13	16
	12	-297	-253	13	14	15
	15	-770	-473	12	14	13
Observed number of days for the H_2O potential to fall to the specified level of -400 kPa →				13	14	
10 August	5	- 39	- 49	17	9	18
Beets	10	- 53	- 70	25	21	26
	14	-100	-117	24	23	24
	18	-212	-191	23	24	22
	21	-311	-234	23	26	23
	25	-501	-370	23	26	23
Observed number of days for the H_2O potential to fall to the specified level of -400 kPa →				23	26	
05 Sept.	2	- 41	- 51	7	6	26
Beets	5	- 47	- 67	15	12	16
	12	-173	-117	18	21	24
	16	-196	-220	23	22	22
	23	-378	-329	25	27	24
	30	-656	-457	27	31	25
Observed number of days for the H_2O potential to fall to the specified level of -500 kPa →				27	32	

(continued)

Table 2. (continued)

22 May	11	- 62	- 60	12	13	-
Beans	14	- 69	- 63	15	16	15
	18	- 80	- 71	17	19	17
	21	-	- 74	-	21	18
Observed number of days for the H_2O potential to fall to the specified level of -75 kPa →				16	20	
01 June	1	- 41	- 47	2	2	44
Beans	4	- 56	- 59	7	6	8
	8	- 66	- 63	12	12	13
	11	- 73	- 68	16	16	18
	15	-111	- 87	17	19	16
	18	-171	-121	17	20	18
	21	-229	-132	18	22	18
Observed number of days for the H_2O potential to fall to the specified level of -150 kPa →				17	22	
02 July	3	-106	- 69	4	5	7
Beans	7	-254	-146	6	8	7
Observed number of days for the H_2O potential to fall to the specified level of -200 kPa →				6	9	
10 July	3	-103	- 66	4	5	10
Beans	7	-298	-141	6	8	8
Observed number of days for the H_2O potential to fall to the specified level of -200 kPa →				5	9	
18 July	2	- 59	- 54	3	4	10
Beans	7	-187	-206	8	8	8
	12	-601	-302	9	12	10
Observed number of days for the H_2O potential to fall to the specified level of -300 kPa →				9	12	
05 August	5	- 48	- 51	7	6	9
Beans	10	- 66	- 64	11	11	11
	14	- 71	- 71	14	14	14
	18	- 71	- 76	19	18	16
Observed number of days for the H_2O potential to fall to the specified level of -75 kPa →				21	17	

The error of one or two days in predicting cycle lengths compares favorably with the errors encountered in scheduling irrigation from daily climatic measurements. Jensen and Wright (1978) show prediction confidence limits of ± 1 day for irrigating alfalfa when the soil water content in the root zone is measured just after the irrigation cycle starts. If the soil water is not measured during the cycle, the confidence limits may be ± several days due to uncertainty of how well the soil profile was wetted.

Ultimately, the uncertainty of all prediction methods must be at least as great as the spatial variation of soil water on a field basis. Jensen and Wright (1978) reported using the neutron meter and measuring soil water to a depth of 75 cm with a standard deviation ranging from 0.7 to 1.1 cm of water. If this range of standard deviation represented the spatial variation in the field, the least uncertainty one might ultimately achieve in predicting irrigation would be ± 1 day, and then only during the midpart of the growing season when transpiration is high. If the soil water was measured gravimetrically as on the plots studied here that had a coefficient of variation of 5.4% and the volumetric water content was 25%, the uncertainty in 75 cm of soil would be 1 cm of water, also giving a minimum uncertainty of at least

± 1 day, and this was a uniform land area. In most practical cases the variability will be greater, indicating there is little to be gained from more accurate individual soil water measurements.

Automation of the gypsum resistance block method offers several potential advantages in predicting irrigation dates when compared to microclimate methods: (1) the block method converges to the correct prediction as time for irrigation nears, (2) it does not require a local crop calibration curve, (3) the amount of water added by irrigation and rain need not be known, (4) the block method appears to be adaptable to some areas where the microclimate approach is difficult to use, such as a shallow water table supplying part of the water for transpiration, and (5) the field truth data could be automatically collected and transmitted from the field making the block method less labor intensive. On the other hand the microclimate approach is well suited for estimating evapotranspiration from large land areas and so is a valuable tool for managing other problems related to soil water evapotranspiration.

A sensor whose resistance is more responsive to water potentials in the -10 to -30 kPa range may be needed in sandy soils. There are also potential instrument problems associated with saline soils that were not studied here. It is possible deeper placement of blocks might be better for some perennial crops having long irrigation cycles due to deep soil and root systems. Nevertheless, the 15-cm depth represents the surface soil zone with the greatest density of crop roots. Most of the nutrients are in this zone and in general, it is this soil volume that must receive optimum management if maximum production is to be achieved. The recommendations for the relatively shallow placement of the blocks as well as the preference for blocks over other soil water instruments for interfacing to a microprocessor are in agreement with the results reported by Shull and Dylla (1980).

CONCLUSIONS

Irrigation dates can be projected using Eq. 1 with gypsum resistance blocks placed in silt loam soil at the 15-cm depth. The accuracy of this method compares favorably with the present scheduling of irrigation from microclimate data. Technology exists to develop a fully automated system. Representative field sites would be instrumented with three to four blocks connected in series to a resistance measuring device that could, upon demand, transmit by wire or radio the resistance to a microprocessor in the manager's office. The microprocessor would interrogate each site daily and store its resistance. Upon demand, this information would be processed through Eqs. 1, 2, and 3 using the linear regression analysis for each irrigation cycle as demonstrated in the last column of Table 2. The only input required by the manager would be an estimate, ± 3°C, of the soil temperature and the water potential at which he wished to irrigate. The microprocessor would keep its own time, referenced to the abrupt decrease in block resistance that occurs during irrigation or rainfall. This type of system should be essentially maintenance free, requiring no labor other than installation of the resistance blocks after planting.

REFERENCES

1. Ben-Asher, J. 1979. Errors in determination of the water content of a trickle irrigated soil volume. Soil Sci. Soc. Am. J. 43:665-668.

2. Cary, J. W. and C. W. Hayden. 1974. Soil strength and porosities associated with cropping sequences. Soil Sci. Soc. of Am. Proc. 38:840-843.

3. Cary, J. W. and W. W. Rasmussen. 1979. Response of three irrigated crops to deep tillage of a semiarid silt loam soil. Soil Sci. Soc. Am. J. 43: 574-577.

4. Cassel, D. K. and L. A. Nelson. 1979. Measurement and statistical analysis of soil water content in field experiments. Agronomy Abstracts, Am. Soc. Agron., Madison, WI. p. 136.

5. Fischbach, P. E. 1978. Basic irrigation scheduling procedures. Irr. Age. April 1978, p. 66 and 70.

6. Gear, R. D., A. S. Dransfield and M. D. Campbell. 1977. Irrigation scheduling with neutron probe. J. of Irr. and Drainage Div., Sept. 1977, Paper No. 13174, p. 291-298.

7. Jensen, M. E. and J. L. Wright. 1978. The role of evapotranspiration models in irrigation scheduling. TRANS. of the ASAE. 21:82-87.

8. Kanemasu, E. T., V. P. Rasmussen and J. Bagley. 1978. Estimating water requirements for corn with a "pocket" calculator. Bul. 615, Agric. Exp. Stat., Kansas State Univ., 24 p.

9. Nielsen, D. R., J. W. Biggar and K. T. Erh. 1973. Spatial variability of field-measured soil water properties. Hilgardia. 42:215-260.

10 Rao, P. V., P. S. C. Rao, J. M. Davidson and L. C. Hammond. 1979. Use of goodness-of-fit tests for characterizing the spatial variability of soil properties. Soil Sci. Soc. Am. J. 43:274-278.

11. Shull, H. and A. S. Dylla. 1980. Irrigation automation with a soil moisture sensing system. TRANS. of the ASAE 23:649-51, 652.

12. Wright J. L. and M. E. Jensen. 1978. Development and evaluation of evapotranspiration models for irrigation scheduling. TRANS. of the ASAE 21:88-91, 96.

SCHEDULING AND MONITORING IRRIGATION WITH THE NEW SOIL MATRIC POTENTIAL SENSOR[1/]

C.J. Phene
Assoc. Member ASAE

J.L. Fouss
Member ASAE

T.A. Howell
Member ASAE

S.H. Patton, M.W. Fisher, J.O. Bratcher, J.L. Rose[2/]

The need to manage irrigation water and the energy required to obtain and distribute that water becomes acute as the availability and cost of these two resources escalate (Phene, 1980). In the arid and semi-arid regions of the world, and particularly in the 17 western states of the U.S., the joint management of energy and water is required in order to sustain the viability and profitability of irrigated Agriculture. Additionally, in these regions, salinity and alkalinity have to be managed along with the irrigation water, and the design and operation of irrigation systems must consider these parameters (Interagency Drainage Program, 1979; Hoffman & Van Genuchten 1981, Van Schilfgaarde et al. 1974; Ayars & Phene, 1981).

Water resource management should be concerned with careful and intensive management of relatively fixed water supplies matching more closely irrigation volumes with crop water requirements in time and magnitude. Internal flexibility to reallocate existing water supplies based on real crop needs rather than acreage must be considered. This type of management will require more site-specific adaptation of irrigation systems and control methods to take full advantage of rainfall, groundwater and waste water in conjunction with soil properties and climate-dependent crop water requirements.

Phene and Howell (1981) demonstrated how to control high-frequency irrigation by achieving an optimal soil matric potential function which limits deep percolation to near zero at the bottom of the root zone. Where needed, salts can be flushed periodically by adjusting the matric potential depth function to permit deep leaching to occur. Frequent irrigation pulses replace water lost to evapotranspiration and the matric potential of a portion of the soil water system is maintained at optimum conditions for aeration and gaseous diffusion. Rawlins and Raats (1975) discussed the implications of high-frequency irrigation on overall soil water balance and crop production. Results reported by Phene and Beale (1976) for experiments conducted under non-saline conditions verify the validity of this method for maintaining nearly constant soil matric potential levels in shallow sandy loam soils. Wu and Gitlin (1981) discuss the statistical implication of the distribution of deficit resulting from system

---

[1/] Contribution from USDA-ARS, CA-HI Area, Water Management Research Laboratory, Fresno, CA 93727, in cooperation with HANCOR, Inc., Findlay, Ohio.

[2/] Supervisory Soil Scientist, USDA-ARS; Vice President, R & D, Hancor, Inc.; Ag Engineer, USDA-ARS; Research Assistant, USDA-ARS, Research Engineer, Hancor, Inc.; Research Assistant and Agronomist, USDA-ARS, respectively.

non-uniformity (hydraulic and plugging) and outline theory for predicting the deficit water supply to a portion of the row when system application uniformity is allowed to decrease.

The successful operation of this automatic control method is greatly dependent upon the accurate measurement or prediction of the soil matric potential of the root zone. Phene et al. (1971a, b) designed a soil matric potential sensor which operated on the principle of heat dissipation from a point source as affected by the amount of water in a ceramic block in equilibrium with soil water. A logic circuit in a feedback or feed-forward mode (Phene et al. 1973) can then initiate or withhold irrigation, based on predetermined soil matric potential set points, and determine the need for irrigation as often as needed by measuring the output of the soil matric potential sensors.

The use of the original sensor was limited by its low voltage output (0-2 mv) within its dynamic range (0-300 kPa). Commercial sensors manufactured by Watertech[3/] (in cooperation with USDA-ARS) circumvented this limitation by linearizing and amplifying the sensor output by a factor of 1000 at the sensing point, thus considerably increasing the signal to noise ratio and permitting remote sensor measurement for agricultural applications. Unfortunately, this modification was costly and required at least six separate electrical conductors to provide power and necessary switching functions in the field. Subsequently Watertech became Moisture Control System, Inc (MCS, Inc., a subsidiary of HANCOR, Inc.) and research was started by MCS, Inc., to simplify the sensor design.

This paper reports a method to monitor and schedule irrigation with the soil matric potential sensor and on the construction changes, testing and use of a simplified soil matric potential sensor developed by MCS, Inc., which utilizes recent technological advances in integrated circuit electronics and a new ceramic material.

CONTROL METHOD

The method for soil sensor control of high-frequency irrigation systems is based on four assumptions: (1) irrigation water is distributed uniformly throughout the field, (2) plant population is uniform in size and distribution, (3) water is used uniformly, mostly by evapotranspiration (E_t), and (4) the soil water sensor is installed in the root zone of an area of the soil-plant-atmosphere system typical of average field conditions or several sensors need to be installed to obtain an average soil matric potential (Phene and Howell, 1981).

If these assumptions are satisfied (i.e., no leaching or runoff and a 100% application efficiency), then theoretically we are dealing with a balance concept which can be described by:

$$I_w = E_t \tag{1}$$

Where E_t represents the total water used from the field by evapotranspiration and I_w is the irrigation water required to compensate for evapotranspiration. Unfortunately, irrigation systems seldom operate at 100% efficiency and we must account for this by applying more water, hence

$$I_w = \frac{E + T}{A_\varepsilon} \tag{2}$$

3/ Trade names are used for identification purposes only and do not imply preference for this item by the U.S. Department of Agriculture.

Where E is soil water evaporation, T is plant transpiration and A_ε is the system's application efficiency ($0 \leq A_\varepsilon \leq 1$). To optimize water use, A_ε can be improved, E can be minimized or T can be decreased by management of the plant canopy, by selecting water efficient plant species, by manipulating planting dates and cultural practices or by application of anti-transpirants. Presently, the best prospect for optimizing water use lies in improving A_ε and decreasing E because decreasing T can be detrimental to crop yields.

Assuming that the 4 basic assumptions are met, then a soil sensor which monitors the soil matric potential in a portion of the root zone can be used to control an irrigation system automatically by replacing water lost from the root zone frequently according to E_t demands (Phene et al. 1979). The system will operate in a simple balance concept, i.e., the water removed from the controlled portion of the root zone will cause the soil matric potential (Ψ_s) to decrease slightly below the preset soil matric potential level (Ψ_p) to cause the conditions $\Psi_s \leq \Psi_p$ which calls for an irrigation pulse. In coarse textured soils, shorter irrigation pulses may be required several times per day to minimize the amplitude of Ψ_s to a narrow range about Ψ_p and maintain steady state conditions. The non-homogeneous differences in soil properties will cause only a slight difference in size and shape of the wetting pattern to which the root system will adjust rapidly as the plant grows. In terms of E_t and plant water stress, frequent irrigations will equalize the differences in soil properties so that everywhere in the field a portion of the root zone will be supplied with enough water to satisfy the average E_t demand.

Van Bavel et al. (1973) simulated water delivery and distribution from an automated subsurface irrigation system (simplified one-dimensional case) controlled by a water content sensor at a given depth which turned irrigation on and off. They concluded that constant water content in the root zone can be achieved provided that the depth of delivery and control are well chosen. Theoretically, if the assumptions are met and the sensor has a rapid response to small changes in Ψ_s, only one sensor in the field is necessary and the exact position of the sensor within the root zone is not critical as long as it is near the source of water. However, in practice, there is some hazard in using only one sensor, in particular, with high-frequency irrigation which does not have much soil-stored water. If the sensor fails and the failure is not detected immediately, then the crop may suffer water stress until the failure is corrected. The effective use of this method has been hampered in the past by the lack of an inexpensive stable electronic sensor capable of operating remotely in a feedback mode with computerized controllers and requiring minimum maintenance during the growing season. The sensor described thereafter fulfills the requirements for use with this method.

CHANGES IN SENSOR CONSTRUCTION

The basic theory of the soil matric potential sensor was described by Phene et al. (1971a) and has remained unchanged. The discrete temperature sensing element (RCA 1N2326, germanium P-N junction diode) has been replaced by a two-terminal integrated circuit temperature transducer (AD 590) which produces an output current proportional to temperature. The AD 590 can be used in any temperature sensing application below +150°C and is available in chip form making it suitable for fast temperature measurements required by the soil matric potential sensor.

The AD 590 is particularly useful in remote temperature measuring applications. The device is insensitive to voltage drops over long lines due to its high impedence current output. Any well-insulated twisted pair of wires is sufficient for operation hundreds of feet from the receiving circuitry.

The original heating element which consisted of 200 cm of teflon-coated copper wire coiled around the diode (R=20 ohms) was replaced by a laser-trimmed hybrid resistor (R=2000 ohms). Increasing the heater resistance by a factor of 100 minimized the effect of long electrical lead resistance on the total voltage drop across the heater.

A new ceramic material developed by Hancor, Inc. (patent applied for) was used to imbed the sensor element. This ceramic material has a pore size distribution capable of providing a large soil matric potential sensitivity between 15 and 90 kPa and some measureable sensitivity between 90 kPa and air dry (3100 kPa). This ceramic provides a stable matrix for the sensor as originally suggested by Phene et al. (1971a) and later used in their modified sensor (Phene et al. 1973).

The water-proofed micro-electronic components are installed in ceramic substrate and imbedded into the ceramic block with thermal conductive polyürethane. Micro electronics allowed the use of a cylindrical ceramic block smaller than that in previous units. The dimensions of the new sensor have thus been reduced to a diameter of 1.4 cm and a length of 2.9 cm.

SENSOR TESTING PROCEDURES

Several characteristics of the sensors were determined to be important for monitoring the matric potential of soil water and for controlling high-frequency irrigation systems (Phene et al. 1971 a, b, 1973). The important characteristics include: (1) high measurement sensitivity in the soil matric potential range of interest for irrigation control and a wide dynamic range, (2) stable calibration with time to insure measurement precision, (3) independence of temperature changes encountered in soils, (4) independence of soil texture and (5) rapid response to changes in soil matric potential. Additionally, it would be advantageous if these sensors could be used with the same calibration curve (i.e., water desorption-voltage output relationship); however, this last requirement, if not met, does not preclude the basic use of the sensor for irrigation control but does require that each sensor be calibrated.

A bridge/amplifier circuit (Fig. 1) was constructed to interface the sensors with a Computer-Based Data Acquisition System (CDAS) (Hewlett Packard, Model 85) (Fig. 2). The Wheatstone Bridge network is similar to that used by Phene et al. 1971. The bridge output is fed into an instrumentation amplifier (LH 0037) followed by a low-bypass active filter (LF 356) to provide a nearly noise-free signal to the CDAS. The instrumentation amplifier amplifies the bridge output by a factor of ten. The heater and the bridge are powered by a constant voltage regulated power supply (HP 6111A) although a battery recharged by a photovaltaic system could be used. The heater current is controlled by the CDAS programmable relay register card in the scanner (HP 3495A). Figure 3 shows the sequence of measurements and calculations performed by the CDAS. Detailed descriptions of computer software and hardware used are available from the authors.

Measurement Range

The maximum dynamic range of 25 sensors was measured by allowing the sensors to slowly air dry until constant sensor measurements were obtained at approximately 3100 kPa (hygroscopic water) (Brady, 1974); then the sensors were saturated in water under 20 kPa vacuum to remove entrapped air. Five air dry and water-saturated measurements were performed at an hour interval each. The mean, standard deviation (SD) and coefficient of variation (CV) of the measurements were calculated.

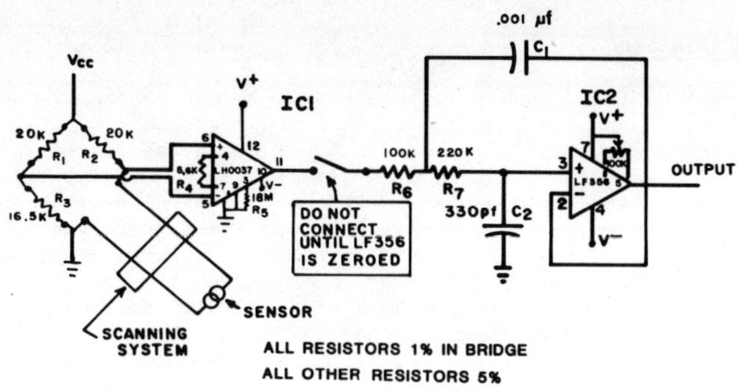

Fig. 1. Bridge/Amplifier/Signal Conditioning Circuit Used to Interface Several Sensors With the Computerized Data Acquisition System.

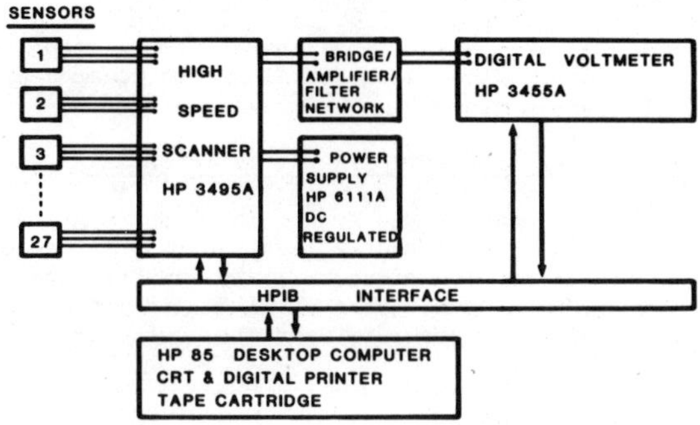

Fig. 2. Computer-Based Data Acquisition System for Sensor Measurements.

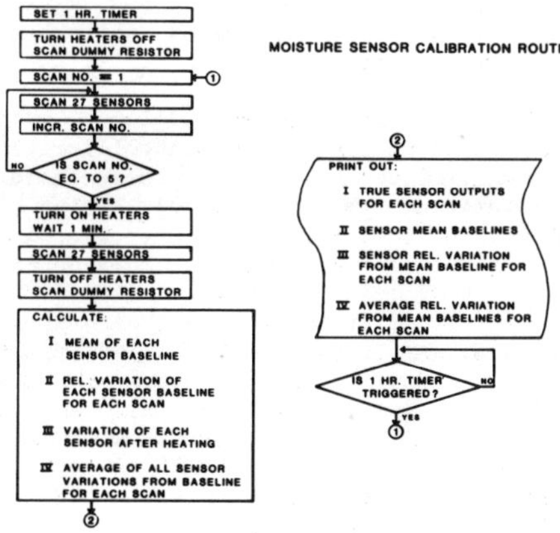

Fig. 3. Measurement and Calculation Sequence Used by the CDAS for Sensor Calibration.

Sensor Calibration

The calibration of the sensor voltage output versus matric potential was accomplished at constant temperature in a pressure plate apparatus (Soil Moisture, Santa Barbara, CA). Each of the sensors was placed in a 300 kPa bubbling pressure ceramic cup (Soil Moisture, Santa Barbara, CA) and covered with a silica powder slurry (200 mesh). Final readings were taken when nearly constant measurements were obtained and there was no water flowing from the pressure plate outflow tube. The last six measurements taken over a 24-hour period were used to determine the mean, standard deviation (SD) and coefficient of variations (CV) of the measurements. The pressure on the plate was then increased. Calibrations were carried out only for a desorption cycle and no attempt was made to study hysteresis.

Temperature Dependence

The temperature dependence of the sensor was tested by equilibrating six sensors at 15 kPa and measuring the sensor output four times at one hour intervals for each temperatures of 5.6, 15.3, 23.2, and 37.0°C. These tests were performed in a constant temperature chamber with the pressure relieved on the pressure plate and the outflow tube clamped after the pressure equilibrium was achieved in the pressure plate.

Equilibrium Time Response to Changes in Matric Potential

The equilibrium time response of the sensor to step changes in matric potential was measured by taking hourly readings of each sensor as the external pressure applied to the pressure plate was increased from one pressure to another. Measurements were obtained and recorded each hour until water equilibrium was reached between the sensor and the applied pressure.

Field Applications of the Method

Several experiments have been conducted in the field using the older version of the sensor to control irrigation (Phene et al. 1973, 1976 and 1981). These various tests demonstrated successfully the validity of the method under extreme dynamic conditions in shallow, layered, sandy loam soils and with potentially high rainfall which had to be accounted for to prevent long periods of water saturated soil.

The soil matric potential sensors were used as input devices for an irrigation controller (HP-85 shown in Fig. 2) to control irrigation in 3 treatments of a replicated irrigation experiment. These field plots were planted to processing tomatoes (Lycopersicon esculentum Mill, UC82B) and irrigated by subsurface trickle irrigation. The objectives of the experiments were to compare methods for scheduling high-frequency subsurface trickle irrigation (direct ET measurements, calculated ET and soil matric potential measurements) with optimized furrow irrigation. In this test, one sensor was measured hourly in each of 3 replicated trickle irrigation plots to provide an average soil matric potential treatment. Originally, three treatments were set to irrigate 4 times daily when the mean soil matric potentials from 4 daily replicated tensiometer readings were respectively -15, -25 and -35 kPa or less at 45 cm depth midway between two emitters and 15 cm from the trickle line; later, after the soil matric potential sensor system was installed, these treatments were changed to: -30, -40 and -55 kPa. Irrigation water applied and yield of the crop grown under these irrigation schedules were compared to that obtained from irrigation scheduling treatments using an open evaporation pan, 75% of open evaporation pan, a screened evaporation pan, evapotranspiration calculated from Van Bavel's equation all modified by a crop factor based on canopy spread, and a furrow irrigated treatement irrigated every 6 days based on optimized commercial practices and previous tomato irrigation trials conducted by the University of California Extension Service.

RESULTS AND DISCUSSION

Measurement Range

The data in Table 1 indicate that the mean voltage output range and standard deviation of 25 randomly selected senors was 255 ± 20 mv and that the coefficient of variation of the mean was 7.73%. When sensors which seem to have matching characteristics were selected from the 25 original sensors and grouped, the mean dynamic range were 246, 254 and 257 mv, respectively, the SD ≤ 12 mv and the CV ≤ 4.8%. Six sensors out of 25 could not be matched with any of the groups or within themselves. For scientific purposes where precision and accuracy of measurements are required, it would be necessary to calibrate each sensor separately unless a 5% CV could be accepted; however, for most commercial irrigation control purposes, the matched groups of sensors could use a single calibration curve using the average of all the sensors. Sensors with similar output calibration curves could easily be identified based on their voltage output range (water saturated and air dry outputs) to be matched in selected groups.

Table 1. Water Saturated Output, Air Dry and Sensor Dynamic Range for all Sensors Grouped Together and for Selected Matched Groups.

# of Sensors	Water saturated output			Air dry output			Output range		
	$\bar{x}$[a]	SD[b]	CV[c]	$\bar{x}$	SD	CV	$\bar{x}$	SD	CV
	- - - mv - - -		%	- - mv - -		%	- - mv - -		%
All Sensors (25)	483	27	5.58	738	38	5.09	255	20	7.73
Selected group of 7 matched sensors	459	8	1.68	705	9	1.24	246	10	4.14
Selected group of 5 matched sensors	475	4	0.81	728	13	1.77	254	12	4.80
Selected group of 7 matched sensors	502	7	1.39	760	11	1.46	257	11	4.20

[a]Mean
[b]Standard Deviation
[c]Coefficient of Variation, $\frac{SD}{\bar{x}}$ X 100

Sensor Calibration

A typical sensor calibration relationship between 10 and 100 kPa is shown in Fig. 4. The data indicate that the sensor has a high sensitivity between 10 and 50 kPa. However, the voltage output range of this sensor (250 mv between 0 and 3100 kPa) indicates that only 56% of the total output is used between 0 and 100 kPa, thus 44% of the voltage range is between 100 and 3100 kPa. Assuming a linear relationship between 10 and 50 kPa the sensitivity is 2.75 mv/kPa, similarly, between 35 and 75 kPa and between

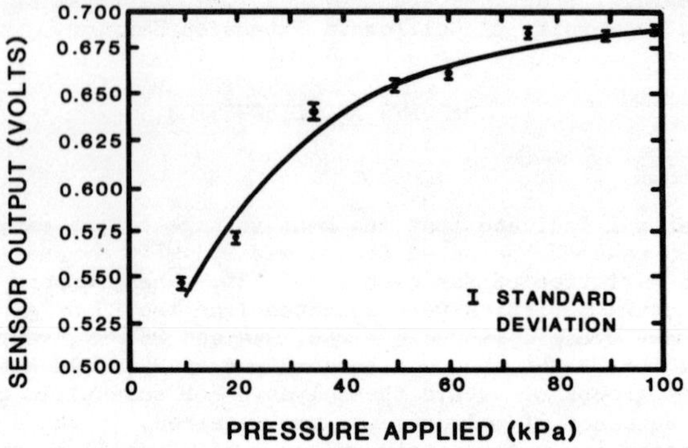

Figure 4. Relationship Between the Sensor Voltage Output and the Pressure Applied on the Pressure Plate.

100 and 3100 kPa the sensitivity is respectively 1.14 mv/kPa and 0.04 mv/kPa. The largest standard deviation is $\pm$ 6 mv @ 35 kPa indicating that the sensor is capable of measuring 35.0 $\pm$ 2.2 kPa at worst in that range. Therefore the sensor meets Criteria 1, i.e., high sensitivity in the soil matric potential range of interest in agriculture and a wide voltage output range.

Time Stability

The sensors have not been used long enough to determine the long-term stability. It is planned, however, that these sensors will be recalibrated after being used in the field for a period of a year to determine whether there was any drift in the original calibration.

Temperature Dependence Test

Results shown in Table 2 indicate that the mean of the 60-sec sensor output at 15 kPa is not temperature dependent over a range of 32°C which the sensors are likely to encounter in agricultural applications. Similar data were obtained for air dry sensors and were previously reported by Phene et al. (1971a). Hence, Criteria 3, i.e., independence of temperature changes encountered in the soil, has been met by the sensor.

Table 2. Temperature Response of 6 Sensors Measured 4 Times Each at One Hour Intervals in a Constant Temperature Chamber.

Temperature (°C)	60 sec. sensor output @ 15 kPa		
	$\bar{x}$[a]	SD[b]	CV[c]
	(mV)	(mV)	(%)
5.6	394.1	0.7	0.18
15.3	378.5	6.9	1.82
23.2	382.3	4.0	1.03
37.0	383.0	2.0	0.53

[a]Mean
[b]Standard Deviation
[c]Coefficient of Variation $\frac{SD}{\bar{x}}$

Independence of Soil Texture

The sensor was tested under water saturated conditions and in air at thermal equilibrium (Phene et al. 1971a). Since the heat pulse travels the fastest in a water saturated sensor, immersing the sensor in water and air provides a broader range of conditions than those likely found in soil. Results of this test performed on six sensors show that there is no difference in voltage output when the wet sensors are immersed in the two mediums. Hence, Criteria 4, i.e., independence of soil texture, has been met by the sensor.

Response to Step Changes in Soil Matric Potential

Figure 5 shows the time required by the sensor to reach water-pressure equilibrium when a step increase in pressure is applied to the pressure plate at several soil matric potentials ranging from 0 to 60 kPa. The time required for equilibrium increases progressively from 1 hr at 10 kPa to 9 hr at 60 kPa and is related to a decrease in the unsaturated conductivity of the ceramic of the sensor and of the calibrating cup. The sensor output cycling shown in Fig. 5 is related to the cycling of the room temperature which causes adiabatic expansion and contraction of the pressure plate equipment. In soil, pressure potential changes occur usually in a continuous mode rather than the discontinous mode created by a step change; therefore, the sensor should respond rapidly to continuous soil matric

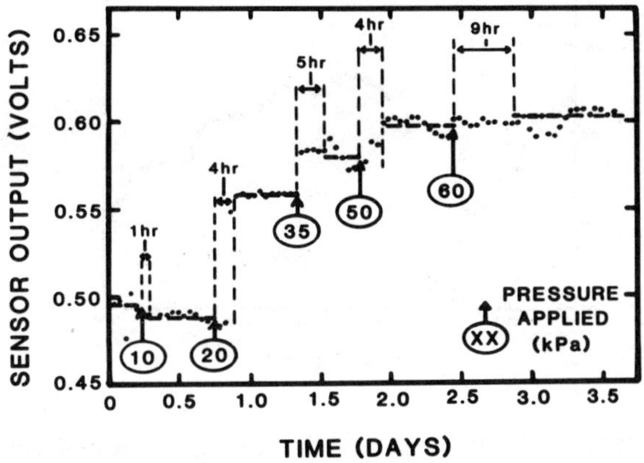

Figure 5. Time Required by the Sensor to Reach Pressure Equilibrium in Response to Steps Increase in Pressure at Different Pressures.

potential changes caused by sinks and sources, provided that the contact between the sensor and the soil is intimate. This is demonstrated by the data in Figs. 6 and 7 which show the response of the sensors to the diurnal variation of the crop E_t. Hence, Criteria 5, i.e., rapid response to changes in soil matric potential, has been met by the sensor.

Field Application of the Method

Figures 6 and 7 show the 45 cm depth soil matric potential fluctuations in experimental field plots irrigated automatically by subsurface trickle irrigation respectively at -30 and -55 kPa or less. This 7 day period followed installation of the sensors in the field and was chosen to demonstrate the concept because the soil was nearly water saturated around the sensor at the beginning and the subsequent measurements demonstrate the

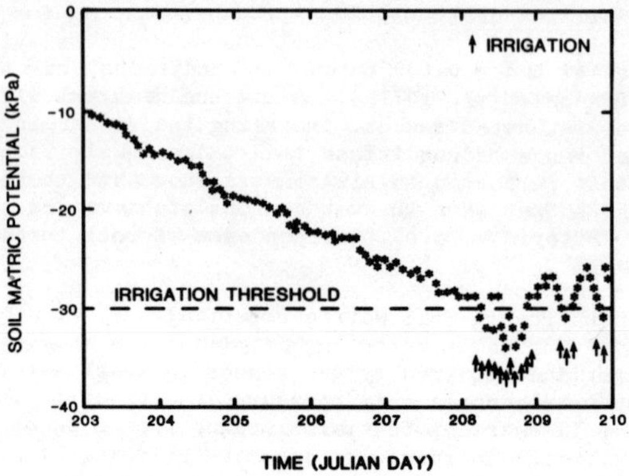

Figure 6. Soil Matric Potential and Irrigation Pulses for Subsurface Trickle Irrigated Tomatoes with Irrigation Threshold at -30 kPa and the Soil Matric Potential Sensor at 45 cm Depth From the Soil Surface (Treatment T6).

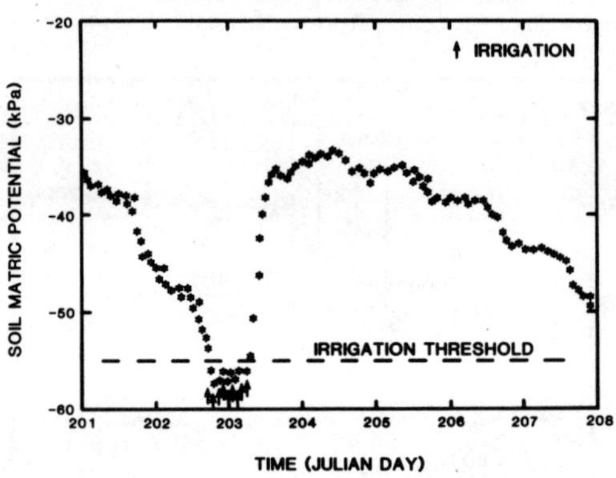

Figure 7. Soil Matric Potential and Irrigation Pulses for Subsurface Trickle Irrigated Tomatoes with Irrigation Threshold at -55 kPa and the Soil Matric Potential at 45 cm Depth From the Soil Surface (Treatment T8).

ability of the sensor to monitor accurately the diurnal soil matric potential fluctuations. The plots were irrigated automatically with 2.5 mm of water when the soil matric potentials in each of treatments T6 and T8 reached -30 and -55 kPa, respectively. Each arrow in Figs. 6 and 7 indicates a 30-min irrigation. During this 7-day period, 16 and 9 irrigations were required to maintain the respective soil matric potentials at -30 and -55 kPa. Eleven irrigations (27.5 mm) and 9 irrigations (22.5 mm) were required to bring the mean soil matric potentials above -30 and -55 kPa respectively. Although the resulting increases in mean soil matric potential were respectively 7 and 23 kPa, the volumetric water content increases were nearly the same. The difference in rooting patterns resulting from previous irrigation treatments will also cause some difference in the water uptake distribution.

The choice of the optimal Ψ_m thresholds (predetermined Ψ_m for initiating irrigation) were based on the average soil characteristics, similarly as those defined by Phene and Beale (1976). The optimal Ψ_m has a range which increases with finer textured soil but is extremely narrow with compacted coarse textured soils. Therefore, Ψ_m characterization of sandy soils is necessary, whereas, one may approximate the optimal Ψ_m for finer textured soils. The choice of Ψ_m = -30 and -55 kPa was based on the soil water desorption characteristics and the relative positions of the sensor and the trickle tube with respect to each other and with respect to the crop rooting pattern. The initial thresholds of -15, -25 and -35 kPa were more adequate for the conditions encountered earlier in the season with a shallower root zone. The threshold Ψ_m's were chosen to provide coverage of the optimal Ψ_m range for the particular soil (i.e., T6 near the wet end of the range, T7 midway and T8 near the dry end of the range). It should be emphasized that with subsurface trickle irrigation it will be necessary to maintain the soil volume around the emitters nearly saturated to prevent roots from penetrating the system. Even though the sensors were not ideally located with respect to the water source to provide a rapid response to irrigation, the soil matric potentials were never lower than 4 kPa below the irrigation threshold levels. In their simulation studies, Van Bavel et al. (1973) have shown that shortening the response time of the system can be achieved by installing the sensors closer to the water source and simultaneously increasing the irrigation threshold levels (irrigating at higher Ψ_m)and decreasing the irrigation volume.

The fruit yields and applied irrigation water are shown in Table 3. Treatments T6, T7 and T8 yielded 125.4, 112.0 and 121.0 t/ha of marketable fruits on 103, 74 and 55 cm of applied irrigation water, respectively. These yields were significantly greater (based on Duncan's separation of means calculation) than that of F1 (furrow irrigated) which was irrigated 11 times with a total of 63 cm of applied irrigation water. Much of the 103 cm of irrigation water was applied to T6 early in the season when irrigation was controlled at -15 kPa based on tensiometer measurements. Since the yield from T8 was 121.0 t/ha (not significantly different from either T6 or T2) with 55 cm of applied irrigation water, the Ψ_m threshold level of -35 kPa early in the season and -55 kPa later in the season could be near optimum for the physical conditions of this specific system. Research is underway to develop a two- and three-dimensional model to define the optimal sensor position - Ψ_m threshold relationship for best control of high-frequency irrigation systems in different textured soils.

SITUATION WARRANTING THE USE OF THIS METHOD

The application of soil matric potential sensor feedback technology may be used beneficially in many agricultural water management applications to improve yield, quality and water use efficiency. Examples of applications are as follows:

Irrigation Method

Trickle irrigation, because of its precise physical design, is more suited for automatic operation than any other methods. The successful use of trickle irrigation or other irrigation methods operating in the high-frequency mode require frequent and precise control of the frequency and volume of irrigation to eliminate the possibility of crop water stress.

Soil Texture

In coarse textured soils, frequent irrigation can increase crop yields, quality and water use efficiency; but as the soil texture becomes finer, the frequency of irrigation may not be as critical (Bucks et al. 1974 and Jury

Table 3. Tomato Fruit Yields and Applied Irrigation Water as Affected by Furrow and Trickle Irrigations and Several Control Methods.

Treatment	Frequency of Irrigation	Marketable Fruit (t/ha)[a]	Non-Marketable (t/ha)[b]	Applied Irrigation Water (cm)
F1 - FI[c]	1/6 days	89.6 d†	15.6 e	63
T2 - TI[d] (open Evap Pan)	2/day	132.2 a	38.0 a	65
T3 - TI (Screen Evap Pan)	2/day	116.4 abc	33.6 ab	61
T4 - TI (75% of T2)	2/day	96.4 c	20.2 d	49
T5 - TI (ET - Van Bavel)	2/day	96.4 c	29.2 abcd	62
T6 - TI (Ψ_m = -30 kPa)	24/day (MAX)	125.4 ab	31.4 abc	103
T7 - TI (Ψ_m = -40 kPa)	24/day (MAX)	112.0 abc	13.4 e	74
T8 - TI (Ψ_m = -55 kPa)	24/day (MAX)	121.0 abc	22.4 bcd	55

[a] Marketable Fruit = Red fruit + 2% green fruit + limited use.
[b] Non-Marketable Fruit = 98% green fruit + blossom end rot + sunscald + small.
[c] FI = Furrow irrigation
[d] TI = Trickle irrigation
† Column means followed by the same letters are not significantly different at the 5% confidence level (Duncan's test on separation of means).

et al. 1977). When fertilizers are injected with the water, frequent irrigation pulses may decrease NO_3 leaching losses and NH_3 volatilization. In soil with low intake rates, high-frequency irrigation can be used to limit application rates so that surface ponding and runoff are minimized or eliminated.

Climatic Environment

In humid and sub-humid areas, the use of automated high-frequency trickle irrigation can minimize water logging problems and thus improve soil aeration because only a portion of the root zone is wetted allowing some storage volume for rainfall. Often in these areas, and particularly in sandy soils, conventional irrigation does not produce the expected yields because excess water impedes aeration and leaches nutrients.

Labor and Safety Factors

In areas where safety is a factor, labor is short and other cost effective operations require automation, the soil sensor feedback technique can provide the simple means for fully automating the irrigation control systems. Soil matric potential and irrigation systems can be monitored from the office via telecommunication systems or radio signals.

Crops

Many permanent (fruit, nut trees and vines) or semi permanent crop (sugar cane) are well suited for trickle irrigation and therefore could benefit from soil sensor feedback control. Because the sensors measure the soil matric potential as affected by the crop E_t during the growing season, the feedback system can provide the flexibility needed to continuously adjust the application volume to match changing the crop ET requirements. With permanently installed subsurface systems as described in this report, the application can be easily adapted for row crops.

Practical Use of this Instrument

This instrument can be used as an input device to any programmable irrigation controller capable of interpreting an analog signal. Although the results of this experiment did not show any yield increase by use of this irrigation control method as compared to the other control methods (direct and calculated E_t) there are certain other advantages in using this instrument:

1. The method can be fully automated and instruments do not neet scheduled maintenance after installation. In case of failure, a warning signal can be provided.

2. The instrument can be sensed remotely via telephone system, direct wiring or radio signal.

3. The instrument and wires can be buried entirely and thus are not subject to vandalism or accident during cultural practices.

4. The measurement will integrate the crop E_t requirements, soil water movement (drainage) and can potentially be used to manage leaching fraction.

Mears et al. (1979) described a prototype of a microprocessor-based irrigation controller capable of receiving inputs from 8 different soil matric potential sensors and controlling 8 solenoid valves. Expansion of the controller to 64 input sensors and solenoid valves could be achieved easily with an optional input/output expander.

This microprocessor-based system was built for under five hundred dollars in parts, although software development costs for the prototype more than offset the cost of hardware.

SUMMARY

A new commercial version of the electronic sensor developed by Phene et al. (1971 a, b) was tested to determine its ability to monitor the matric potential of soil water and to control high-frequency irrigation systems. Several important characteristics were evaluated, and it was found that these sensors have (1) a large measurement sensitivity in the soil matric potential range of interest for irrigation control and a wide voltage output range, (2) independence of temperature changes encountered in soils, (3) independence of soil texture and (4) a rapid response to changes in soil matric potential.

A commercial CDAS was adapted to monitor the soil matric potential with these sensors and to determine the need for irrigation. Software to calibrate the sensors and to control automated irrigation systems were developed and were tested in a replicated field irrigation study. Results demonstrated that high-frequency trickle irrigation can be controlled accurately by soil matric potential sensors in the root zone of row crops and produce yield of processing tomatoes comparable to that obtained with E_t-based control methods. The sensors can be used in a feedback mode to maintain a portion of the root zone at nearly constant Ψ_s within an optimal Ψ_s range.

ACKNOWLEDGEMENT

The authors express their appreciation to the Plant Science Department at California State University, Fresno where the field test was conducted and to R.D. Beck, R.L. Chavez, S. Ballew, D. Robb and the support staff at the Water Management Research Laboratory for their participation in this test.

REFERENCES

1. Ayars, J.E. and C.J. Phene. 1981. Managing Irrigation Systems in the San Joaquin Valley. ASAE Paper No. 81-2099.

2. Brady, N.C. 1974. The Nature and Property of Soils. McMillan Publishing Co., Inc. New York, N.Y. pp. 191.

3. Bucks, D.A., L.J. Eriee and O.F. French. 1974. Quantity and Frequency of Trickle and Furrow Irrigation for Efficient Cabbage Production. Agronomy Journal 66: 53-57.

4. Hoffman, G.J. and M.TH. van Genuchten. 1981. Efficient Water Management for Salinity Control. In A.S.A. Monograph Limitations to Efficient Water Use in Crop Production, Chap. 2C, Eds. H.M. Taylor, W.R. Jordan, T.R. Sinclair (In Press).

5. Interagency Drainage Program. 1979. Agricultural Drainage and Salt Management in the San Joaquin Valley, Final Report. California Dept. of Water Resources, U.S. Bureau of Reclamation, California State Water Resources Control Board. June.

6. Jury, W.A. and K.D. Earl. 1977. Water Movement in Bare and Cropped Soils Under Isolated Trickle Emitters: I. Analysis of Bare Soil Experiment. Soil Sci. Soc. American Journal 41: 852-856.

7. Mears, D.B., C.J. Phene, and S.D. Soenke. 1979. Microprocessor-Based Irrigation Controller. ASAE Paper No. 79-5525.

8. Phene, C.J., G.J. Hoffman, and S.L. Rawlins. 1971(a). Measuring Soil Matric Potential In Situ by Sensing Heat Dissipation within a Porous Body: I. Theory and Sensor Construction. Soil. Sci. Soc. Am. Proc. 35: 27-33.

9. Phene, C.J., S.L. Rawlins, and G.J. Hoffman. 1971(b). Measuring Soil Matric Potention In Situ by Sensing Heat Dissipation within a Porous Body: II. Experimental Results. Soil Sci. Soc. Am. Proc. 35: 225-229.

10. Phene, C.J., G.J. Hoffman, and R.S. Austin. 1973. Controlling Automated Irrigation with Soil Matric Potential Sensor. Trans. ASAE 16: 773-776.

11. Phene, C.J. and O.W. Beale. 1976. High-Frequency Irrigation for Water Nutrient Management in Humid Regions. Soil Sci. Soc. Am. J. 40: 430-436

12. Phene, C.J., J.L. Fouss and D.C. Sanders. 1979. Water-Nutrient Management of Potatoes with Trickle Irrigation. Am. Potato Journal 56: 51-59.

13. Phene, C.J. 1980. Contraints Face Irrigated Agriculture. California Arizona Farm Press, Vol. 2, No. 43, pg. 30, 32-34, 38, 40. Oct.

14. Phene, C.J. and T.A. Howell. 1981. Control of High-Frequency Irrigation System. ASAE Paper 81-2013.

15. Rawlins, S.L. and P.A.C. Raats. 1975. Prospects for High-Frequency Irrigation. Science 188: 604-610.

16. Van Bavel, C.H.M., J. Ahmed, S.I. Bhuiyan, E.A. Hiler and A.G. Smajstrla. 1973. Dynamic Simulation of Automated Subsurface Irrigation Systems. Trans. ASAE 16: 1095-1099.

17. van Schilfgaarde, J., L. Bernstein, J.D. Rhoades, and S.L. Rawlins. 1974. Irrigation Management for Salt Control. Journal of Irrigation and Drainage Division. Amer. Soc. of Civ. Engineers. Vol. 100, No. IR 3, pg. 321-338. September.

18. Wu, I-Pai and H.M. Gitlin. 1981. Drip Irrigation Application Efficiency and Schedules. ASAE Paper No. 81-2078.

SCHEDULING AND CONTROLLING IRRIGATIONS FROM MEASUREMENTS OF SOIL ELECTRICAL CONDUCTIVITY

J. D. Rhoades, D. L. Corwin, G. J. Hoffman
Member ASAE

ABSTRACT

When salinity is significant, measurements of soil water content or matric potential are not sufficient for scheduling irrigations nor do they permit control of leaching fraction. In this paper the concept and principle of using measurements of bulk soil electrical conductivity, σ_a, to schedule irrigations, control the depth of water penetration and obtain a desired leaching fraction are introduced and supporting evidence is presented. It is suggested that some irrigation systems could be automated with burial-type four electrode probes to initiate an irrigation when σ_a at a shallow depth declines to a predetermined set point value and to terminate when the wetting front arrives at a desired lower depth sensor location and that irrigation amount can be adjusted over time to achieve a desired leaching fraction from feedback information of σ_a in the lower region of the rootzone.

INTRODUCTION

Practical means of scheduling and controlling irrigations to conserve water while avoiding yield loss from salinity or water deficiency are lacking. Irrigation scheduling methods based solely on measurements of soil water content or matric potential are inadequate where salinity is a hazard because of the added osmotic stress which in turn, also depends upon soil water depletion between irrigations. Furthermore, direct measurements of soil water content or matric potential can not be used to control the leaching fraction (the fraction of infiltrated water that passes the rootzone) which is required to prevent excessive soil salinity accumulation. For saline water, irrigations must be scheduled before the total soil water potential (matric plus osmotic) drops below the level which permits the crop to extract sufficient water to sustain its physiologic processes without loss in yield. The crop's root system normally extracts progressively less water with increasing soil depth because rooting density decreases with depth and because available soil water decreases with depth as salt concentration increases (Rhoades and Merrill, 1976). Therefore, the frequency of irrigations should be determined by the total soil water potential in the upper rootzone where the rate of water depletion is greatest. On the other hand, the amount of water to apply depends on stage of plant development and the salt tolerance of the crop and, consequently, should be based on the status of the soil water at deeper depths. In early stages of plant development it is often desirable to irrigate to bring the soil to "field capacity" to the depth of present rooting or just beyond. Eventually, however, water must be applied to leach out some of the salts accumulated in the profile to prevent salt concentrations from exceeding tolerable levels. Thus, the amount of water required is dictated by volume of soil reservoir in need of replenishment and level of soil salinity in the lower rootzone.

The Authors Are: J. D. RHOADES, Research Leader, and D. L. CORWIN, Soil Scientist, and G. J. HOFFMAN, Research Leader, U. S. Salinity Laboratory, 4500 Glenwood Drive, Riverside, CA

Soil water salinity (and hence osmotic potential) and leaching fraction can be determined from measurements of soil electrical conductivity, σ_a (Rhoades, 1980/81). σ_a is responsive to both soil water content and soil salinity. Soil salinity in irrigated agriculture is normally low at shallow soil depth and increases through the rootzone. Thus measurements of σ_a shallow in the soil profile made over an irrigation cycle are relatively more indicative of changing soil water content while measurements deeper in the profile, where little water uptake occurs, are more indicative of salinity. Both soil water depletion and depth of water penetration from an irrigation or rainfall can be determined from σ_a measurements made within the rootzone over time. Automated readings of σ_a are feasible using burial type 4-probe sensors (Rhoades, 1979). Achieving highly efficient irrigated agriculture should be facilitated from such periodic "feedback" information on the status of salt and water within the crop rootzone and the extent of leaching being achieved.

This paper describes the concept and theory of using measurements of σ_a for scheduling and controlling irrigations and gives preliminary results in support of the concept. Instruments and techniques for measuring σ_a have been described elsewhere (Rhoades, 1978).

Principles of Soil Electrical Conductivity

The relationship between the bulk soil electrical conductivity, σ_a; the electrical conductivity of the entrained soil water, σ_w; the volumetric content of soil water, θ, and the matrix conductivity, σ_s; was previously presented by Rhoades, et al. (1976):

$$\sigma_a = [\sigma_w \, \theta] \, T + \sigma_s \qquad [1]$$

T is an empirically determined "transmission" coefficient (or reciprocal tortuosity factor) which is linearly related to θ by

$$T = a \, \theta + b, \qquad [2]$$

with constants a and b determined by regression analysis. Both T and σ_s are properties of the soil solid phase, while σ_w and θ are properties of the soil liquid phase.

Scheduling Irrigations

After rapid drainage ceases following an irrigation and the soil is at "field capacity," further major losses of soil water occur by evapotranspiration. Almost all of the salts are excluded from the plant by the root membranes and no salt leaves by evaporation; hence the salt concentration (or electrical conductivity) of the remaining soil water is increased as θ is reduced by evapotranspiration. Because of this inverse proportional relationship between σ_w and θ, the product $(\sigma_w \, \theta]$ found within a given soil volume at field capacity will not change appreciably as θ is reduced, so $\sigma_w \cdot \theta \simeq K$ (a constant). Changes in θ do affect σ_a, however, through its influence on T (see Eq.[2]). Thus, as θ decreases below field capacity by evapotranspiration σ_a will decrease approximately linearly by the relationship

$$\sigma_a = [K](a\theta + b) + \sigma_s \qquad [3]$$

Therefore, at a given shallow soil depth, a σ_a value should exist that is characteristic of the onset of plant stress. This value can be used as a set point value (σ_a') for scheduling irrigations. The actual value of σ_a' will vary with soil type, irrigation water quality and crop salinity, but it is a characteristic value that is analogous to tensiometer or neutron probe readings used to initiate an irrigation. It is possible to predict such σ_a' values, but this will be the subject of a subsequent paper.

Replenishing Soil Water

At the end of an irrigation cycle, a certain σ_a-depth relation will exist through the soil profile. Upon irrigation, the water content will increase at every depth in the profile reached by the wetting front and σ_a will increase correspondingly as water of σ_w flows into that soil volume and θ increases. Increases in σ_a observed successively with depth and time following an irrigation are evidence of replenishment of soil water at these depths. Thus, irrigation amount can, in principle, be controlled to achieve replenishment to any desired depth.

Controlling Leaching Fraction

The principle of determining leaching fraction, L, from measurements of bulk soil electrical conductivity has been presented previously (Rhoades, 1980/81). Briefly, at or below the bottom of the rootzone σ_w is directly related to (1/L) when θ is at field capacity. Thus, (1/L) and σ_a are linearly related as

$$(1/L) = c\,\sigma_a + d \qquad [4]$$

Thus, irrigation amounts for a given soil could, in principle, be adjusted over a period of time to achieve the desired L.

EXPERIMENTAL PROCEDURE

Four electrode salinity probes (see Fig. 1) were installed at five soil depths (15, 30, 60, 90, and 120 cm) in the rhizosphere plots at the U. S. Salinity Laboratory where an irrigation management study was in progress (Fig. 2).

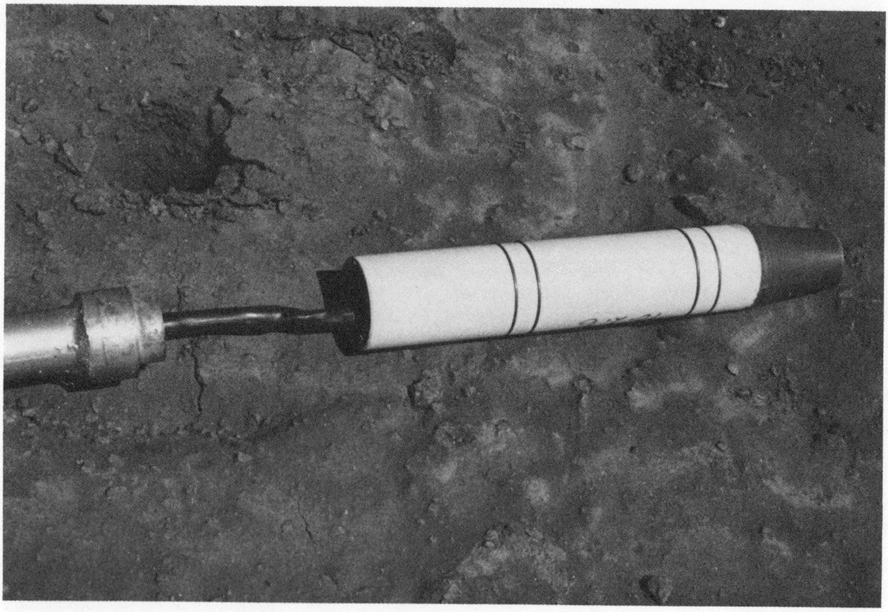

Fig. 1. Four-electrode burial probe.

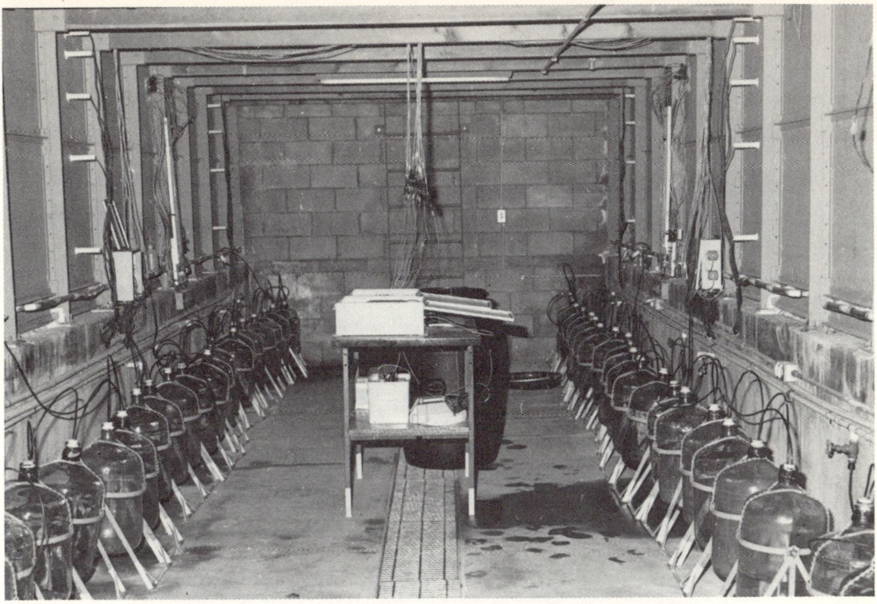

Fig. 2. View of rhizosphere plots from below showing four-electrode probes inserted into the soil profile and wires leading to data acquisition system.

These 3 by 3 by 1.5 m deep plots were filled with Pachappa fine sandy loam top soil and equipped to permit controlled irrigations on a variable frequency basis with waters of different salinity levels and at different leaching fractions. Neutron meter access tubes were installed in each plot to monitor soil water content. Irrigations of the tall fescue crop were scheduled based upon neutron-probe readings, pan evaporation, and the ratio of drainage to irrigation volumes. σ_a measurements were made throughout several irrigation cycles using a specially designed data-logging system (Fig. 3).

Fig. 3. Data acquisition system used to make and record soil electrical conductivity readings.

RESULTS

The rapid responses of 4-probe sensors to changes in soil water content and salinity following an irrigation are illustrated in Figs. 4, 5, and 6.

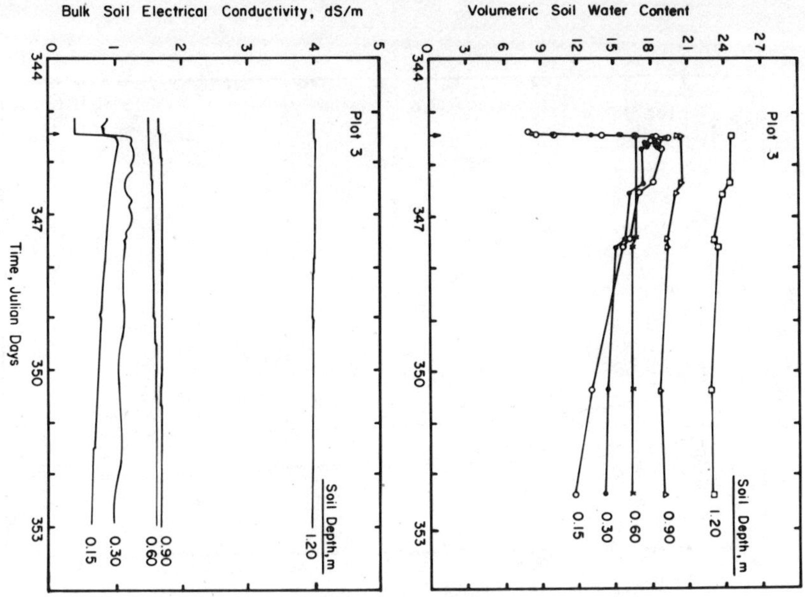

Fig. 4. Measurements of water penetration following a small irrigation with 4-probe sensors and a neutron probe.

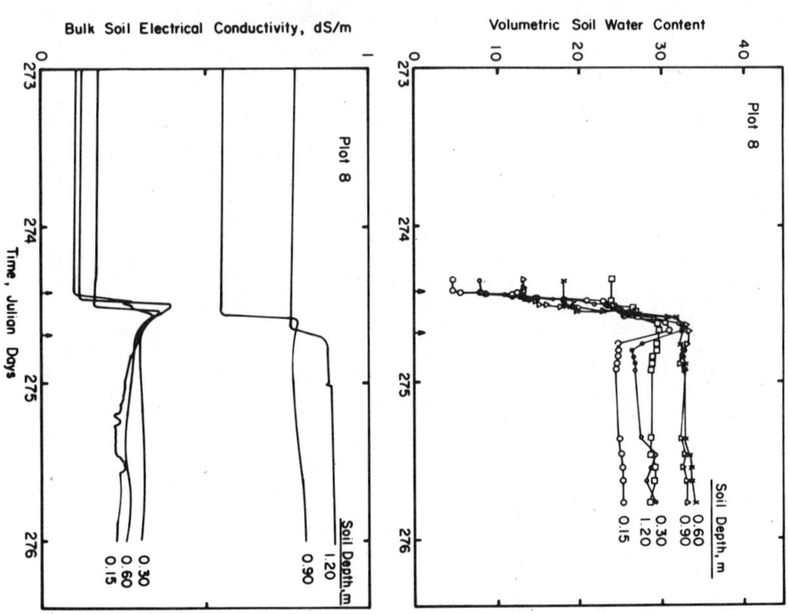

Fig. 5. Measurements of water penetration following a large irrigation with 4-probe sensors and a neutron probe.

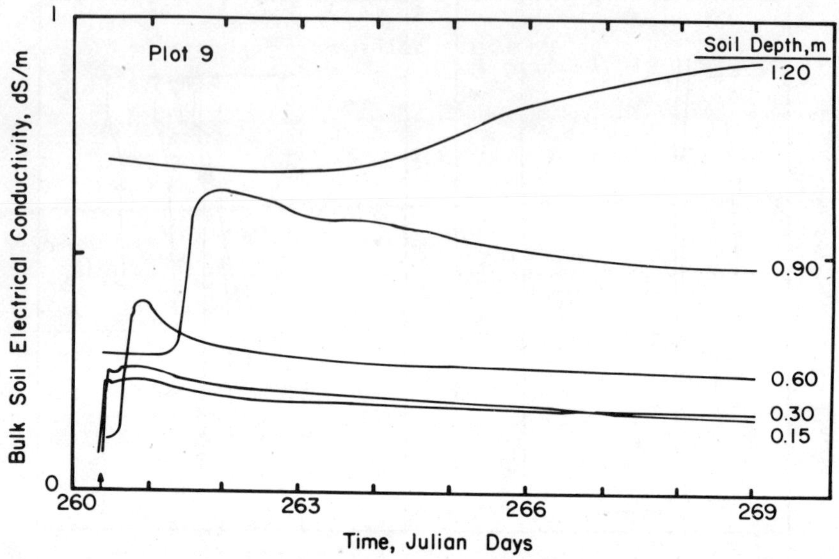

Fig. 6. Measurement of water penetration following an intermediate irrigation with 4-probe sensors.

This sensitivity demonstrates the feasibility of determining the depth of water replenishment by σ_a sensors. Fig. 4 summarizes σ_a and soil water content measurements prior to, during, and following a small irrigation of 23 mm applied in 44 min., an average application rate of 750 mm per day. The arrow on the abscissa indicates the brief irrigation period. The σ_a readings show that the irrigation water arrived at the 15 and 30 cm soil depths as a sharp front; the front did not reach the 60 cm depth although some water reached 60 cm as indicated by the slight but persistent increase in σ_a during the week following the irrigation. The constant σ_a readings at 90 and 120 cm depths indicate that no water penetrated to these deeper depths. Volumetric soil water content measured with a neutron probe, also shown in Fig. 4, confirms the water movement pattern measured with the σ_a sensors. Sensor response to a large irrigation of 254 mm applied in 6.4 hrs. is shown in Fig. 5. The σ_a readings show that this irrigation caused water to move through the entire 1.2-m-deep soil profile. σ_a readings changed drastically as the wetting front passed each sensor at the different depths with the wetting front reaching a depth of 1.2 m about 5 hrs. after irrigation was initiated. Soil water content values measured with the neutron meter (Fig. 5) support the arrival of the wetting front at each soil depth with the maximum soil water content at the 1.2 m depth being reached about 5 hrs. after irrigation began. An example of an irrigation causing an intermediate depth of soil water replenishment is given in Fig. 6. A total of 173 mm was applied with an initial application of 73 mm applied over a period of 148 min.; the remainder was applied in twenty, 10-min. pulses. The σ_a data show the progressive advance of the wetting front to the 90-cm soil depth and then drainage from the shallow profile down to the 120-cm depth. These three examples demonstrate the feasability of using σ_a measurements to control irrigation amount to achieve soil water replenishment to a desired depth in the soil profile.

Following an irrigation, the soil profile quickly drains to "field capacity" as shown, for example, in Fig. 5. Likewise, σ_a, see Fig. 5, reaches a steady value shortly after irrigation ceases and then remains relatively constant for several days. These data demonstrate that soil salinity can be determined from σ_a measurements for several days following an irrigation as hypothesized earlier (Rhoades and Ingvalson, 1971).

Representative σ_a data obtained during several irrigation cycles are presented in Fig. 7. As evident in these figures, σ_a at shallow depths comes to reproducible and characteristic values just prior to each irrigation as hypothesized. Hence, irrigations could be scheduled based upon shallow-depth σ_a readings.

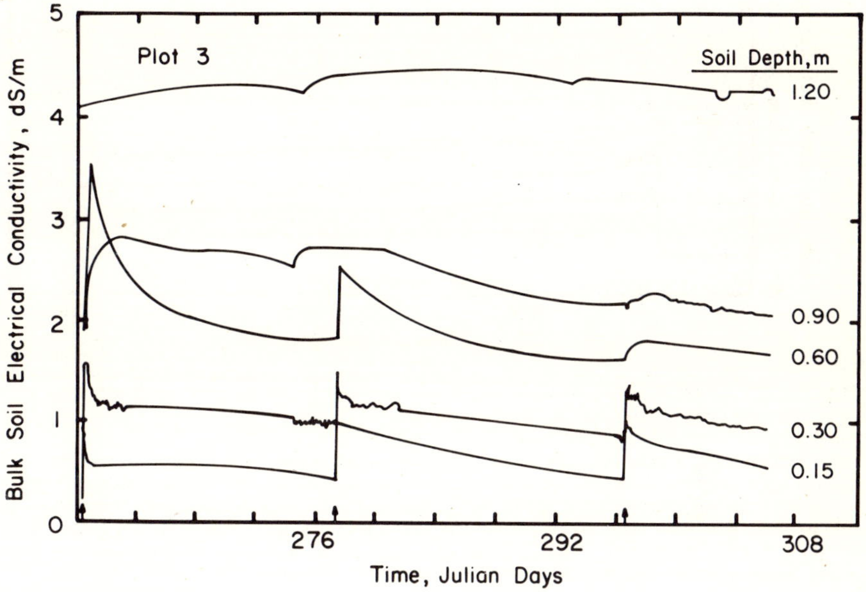

Fig. 7. Measurements of soil electrical conductivity over several irrigation cycles.

Data from a portion of one irrigation cycle shown in Fig. 7 are presented in Fig. 8 in detail.

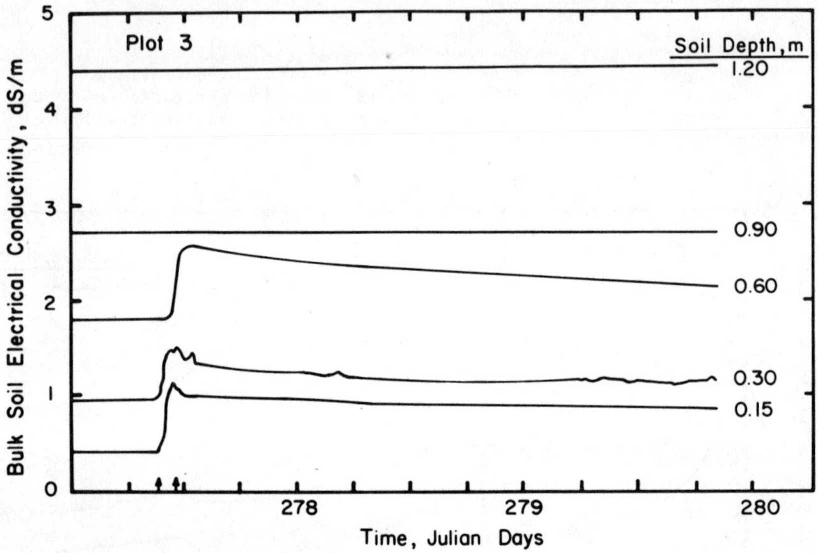

Fig. 8. Measurements of soil electrical conductivity over a several day period following an irrigation.

The deep sensor readings are nearly steady with time because of the near constancy of water content and solute concentration. As explained earlier, σ_a at the lower extremity of the rootzone should be inversely proportional to leaching fraction. The soil in the rhizosphere is uniform with depth. Thus it is apparent from Eq. (1) that immediately following an irrigation the ratio of σ_a values near the top and at the bottom of the rootzone is equivalent to leaching fraction since the corresponding values of σ_w are those of the irrigation (σ_i) and drainage (σ_d) waters, respectively, and $L = \sigma_i/\sigma_d$. In this particular case, the leaching fraction determined from the σ_a ratios is 0.20 (0.88/4.46) while that measured from σ_i/σ_d was 0.21. These data support the hypothesis that irrigations could potentially be controlled from measurements of σ_a in the lower rootzone to achieve desired leaching fractions.

CONCLUSION

The data presented here support the hypotheses that measurements of bulk soil electrical conductivity could be used to schedule irrigations, control the depth of water penetration, and obtain a desired leaching fraction. Some irrigation systems could be automated with burial-type four electrode probes to initiate an irrigation when the σ_a at a shallow depth declines to a predetermined appropriate set point value and to terminate when the wetting front arrives at a desired lower depth sensor location. One should be able to adjust irrigation amount over time to achieve a desired leaching fraction from feedback information of σ_a in the lower region of the rootzone.

REFERENCES

1 Rhoades, J. D., and R. D. Ingvalson. 1971. Determining salinity in field soils with soil resistance measurements. Soil Sci. Soc. Am. Proc. 35: 54-60.

2 Rhoades, J. D., and S. D. Merrill. 1976. Assessing the suitability of water for irrigation: Theoretical and empirical approaches. FAO Soils Bulletin 31: 69-109.

3 Rhoades, J. D., and J. van Schilfgaarde. 1976. An electrical conductivity probe for determining soil salinity. Soil Sci. Soc. Am. J. 40: 647-651.

4 Rhoades, J. D., P. A. C. Raats, and R. J. Prather. 1976. Effects of liquid-phase electrical conductivity. Soil Sci. Soc. Am. J. 40: 651-655.

5 Rhoades, J. D. 1979. Inexpensive four-electrode probe for monitoring soil salinity. Soil Sci. Soc. Am. J. 43: 817-818.

6 Rhoades, J. D. 1978. Monitoring soil salinity: a review of methods. In L. G. Everett and K. D. Schmidt (eds.), Establishment of Water Quality Monitoring Programs, Am. Watr. Resources Assoc., San Francisco, June 3-8, 1978: p. 150-165.

7 Rhoades, J. D. 1980/81. Determining leaching fraction from field measurements of soil electrical conductivity. Agr. Water Mgt. 3: 205-215.

IRRIGATION SCHEDULING IN SUBHUMID AREAS WITH INFRARED THERMOMETRY[1/]

D. C. Slack
Member ASAE

K. M. Geiser
Assoc. Member ASAE

K. W. Stange
Assoc. Member ASAE

E. R. Allred
Member ASAE

During the past five years, irrigated acreage in the eight midwestern states, east of the Dakotas and Nebraska, has more than doubled to over 800,000 ha in 1981 (Anon. 1980). While this total is relatively low compared to regions where irrigation has long been practiced, the increase is substantial. Proper timing of irrigation water application in this subhumid region presents some challenges and opportunities peculiar to the region. First, rainfall is an important element in the crop-water supply and must be accounted for. Secondly, climatic factors affecting crop-water use tend to vary significantly over the growing season making it necessary to calibrate simple evapotranspiration (ET) models for such conditions. Finally, the acreage in any one area is often small enough that the demand for services is not sufficient to support irrigation scheduling consultants. Thus, there is considerable need for a farmer-oriented scheduling method for this region.

Plant temperature has long been recognized as an indicator of crop-water stress (Gates 1964; Hiler et al. 1974). However, it has been only with the advent of small, hand-held, relatively inexpensive infrared thermometers that routine measurement of canopy temperature as an irrigation scheduling tool has become practical. Some of the initial work with these devices indicated that the canopy temperature (T_c), or canopy-air temperature difference (ΔT), may be a sufficient indicator of crop-water stress for scheduling purposes (Jackson et al. 1977; Idso et al. 1977). However, more recent work has shown that a number of factors influence ΔT and that these factors must be accounted for if ΔT is to be used as a reliable irrigation scheduling tool (Geiser et al. 1980; Jackson et al. 1981).

The purpose of this paper is to:

(1) discuss the factors influencing ΔT,

(2) describe a ΔT based irrigation scheduling procedure utilizing infrared thermometry, and

(3) present some results of an experimental evaluation of this procedure.

1/ Scientific Journal Series No. 11,836 of the Minnesota Agricultural Experiment Station.

The authors are: D. C. Slack, Associate Professor, Agricultural Engineering Department, University of Minnesota, St. Paul, MN; K. M. Geiser, Engineer, Johnson Division U.O.P., St. Paul, MN; K. W. Stange, Instructor, Agricultural Engineering Department, South Dakota State University, Brookings, SD; and E. R. Allred, Professor, Agricultural Engineering Department, University of Minnesota, St. Paul, MN.

THEORETICAL BASIS

The temperature of a crop canopy indicates the state at which the energy exchange between the crop and the surrounding environment is in equilibrium. Thus, changes in convection, radiation and transpiration affect the canopy temperature. A simple energy balance for the canopy can be written as:

$$R_n = H + \lambda E + G \tag{1}$$

where R_n is the net radiation ($W \cdot m^{-2}$), H is the sensible heat flux ($W \cdot m^{-2}$), λE is the latent heat flux ($W \cdot m^{-2}$), G is the heat flux below the canopy ($W \cdot m^{-2}$) and λ is the heat of vaporization of water ($J \cdot kg^{-1}$).

Using a gradient-resistance form of equation, the terms H and λE may be expressed as:

$$H = \rho C_p (T_c - T_a)/r_a \tag{2}$$

and

$$\lambda E = \lambda \rho (q_c - q_a)/r_1 \tag{3}$$

Where ρ is the density of the air ($kg \cdot m^{-3}$), C_p the specific heat capacity of the air ($J \cdot kg^{-1}$), T_c the canopy temperature (°C), T_a the air temperature (°C), r_a is the aerial diffusion resistance to heat transfer ($s \cdot m^{-1}$), q_c the concentration of water vapor at the canopy surface at temperature T_c, q_a the water vapor concentration in the air at temperature T_a and r_1 is the resistance to vapor transfer ($s \cdot m^{-1}$).

Substituting Eqs. (2) and (3) into Eq. (1), assuming that G is negligible (Decker 1964) and solving for $(T_c - T_a)$ yields:

$$\Delta T = T_c - T_a = \frac{r_a}{C_p} \left[\frac{R_n}{\rho} - \lambda\left(\frac{q_c - q_a}{r_1}\right)\right] \tag{4}$$

Equation (4) illustrates that ΔT is influenced by a number of factors. Obviously net radiation has a direct effect. However, if net radiation and other environmental factors are assumed to remain constant for a short time period, ΔT may be seen as responding primarily to the water-vapor flux. This flux is controlled by the gradient in water-vapor concentration between the plant and the atmosphere, and the vapor resistance term. If q_c is assumed to be that at saturation, and q_a is constant, then the mechanism controlling vapor flux is the canopy resistance r_1 which is dominated by the stomatal diffusion resistance r_s ($s \cdot m^{-1}$). While numerous factors affect stomatal action, plant-water stress and incident radiation are the two most prominent (Turner 1974; Sumayao 1979). The influence of incident radiation on stomatal opening can be overridden by the degree of plant-water stress (Ehrler and van Bavel 1967). Thus, stomatal diffusion resistance has gained acceptance as an indicator of crop-water stress, and in measuring ΔT we are making an indirect measurement of this term providing other factors remain constant. Stange et al. (1981) found that for a relatively high level of net radiation, stomatal closure on corn under field conditions occurred when ΔT approached +3°C. It was also found that the stomatal diffusion resistance at this point corresponded closely to that at a "critical" or "threshold" value of leaf-water potential of about -15 bars which also indicated the onset of stomatal closure. Thus, ΔT appears to be a good measurable indicator of crop stress under relatively constant environmental conditions. However, Eq. (4) indicates that as environmental factors change the critical ΔT indicating stomatal closure would also change.

Some investigators have proposed scheduling methods which eliminate environmental effects by comparing the ΔT of the crop to be scheduled with that of a well watered crop. Clawson and Blad (1980) successfully used such an approach to schedule irrigations on corn. The usefulness of such an approach for a farmer-oriented scheduling method, however, is questionable. A more useful variation of this approach is to use the variability of canopy temperature within a field to indicate plant-water stress. Blad et al. (1980) concluded that a standard deviation of mid-day canopy temperature exceeding 0.3°C indicates stress in some parts of a field.

Studies by Jackson et al. (1977) and Idso et al. (1977) have indicated that wheat is not stressed for water unless leaf temperature exceeds air temperature. However, Reicosky et al. (1980) concluded that at mid-day the temperature of fully exposed leaves near the surface of a well-watered canopy will be slightly below air temperature in arid or semi-arid climates and slightly above air temperature in humid or subhumid climates.

DEVELOPMENT OF A SCHEDULING PROCEDURE

There are two important elements to a successful irrigation scheduling procedure. First, an indication is needed of when to irrigate, and secondly, it is necessary to know how much water to apply. Most soil-moisture-based scheduling methods use the soil to determine both criteria (e.g., irrigate when available water is 50% depleted). Since an objective of irrigation sheduling is to minimize plant-water stress, it would seem desirable to use the plant as an indicator of when to irrigate and the soil to determine how much to apply. This suggests that measurement of both plant and soil parameters is required for such a system. However, the effect of plant-water potential on stomatal closure has been fairly well established for a number of crops (Stange et al. 1981; Turner 1974; Beadel et al. 1973; Kanemasu and Tanner 1969). Plant-water potential is in turn closely related to soil-water potential (Ψ_s) (bars) or available soil-water (AW) (%). Thus, since it seemed that a relationship could be developed relating ΔT to soil-water status and other environmental factors, it was decided to use Eq. (4) as the basis for developing such a relationship by regression techniques.

Referring to Eq. (4), it is seen that ΔT is a direct function of R_n and the gradient in water vapor concentration ($q_c - q_a$) or (Δq). If the site of evaporation at the leaf surface is considered to be saturated, Δq can be expressed as a function of q_a or atmospheric relative humidity (RH). The aerodynamic resistance (r_{al}) is a function of leaf shape, orientation, temperature and wind speed $W(m \cdot s^{-1})$. For a given crop and a narrow range of temperature, r_a should vary primarily with wind speed. The remaining variable, r_1, is dominated by the stomatal diffusion resistance (r_s), which is directly related to plant-water potential (Ψ_p) (bars). Thus, it should be possible to develop a regression equation with the following variables:

$$\Delta T = f(R_n, RH, W, AW) \quad (5)$$

or

$$\Delta T = f(R_n, RH, W, \Psi_s) \quad (6)$$

To develop the regression equation of the form of Eq. (5), data were collected on field corn subjected to various levels of stress over three growing seasons (1978-80) at the University of Minnesota Sand Plains Experiment Station. The data collected included ΔT as measured with an infrared thermometer positioned 1 m above the canopy with a viewing angle 30° downward from the horizontal using the configuration described by Grannes et al. (1977) and Geiser (1981). R_n was measured with a Fritschen type net radiometer positioned 0.5 m above the canopy and level. AW was measured with a 3-cup anemometer 0.5 m above the canopy. AW was determined using a neutron probe. A more detailed description of the experimental procedure is given by Geiser et al. (1980) and Geiser (1981). Regression analysis of data collected during the 1978 and 1979 growing seasons indicated that wind speed was not a significant variable even when treated as an interactive term with R_n or RH. Carlson et al. (1972) and Barthekar (1975) found similar results in relating ΔT to W. Barthekur (1975) also found that above a velocity of about $2 m \cdot s^{-1}$, increases in wind speed had a small affect on leaf temperatures. Other variables were significant and step-wise regression techniques yielded the following relationship:

$$\Delta T = -1.065 + 4.71x10^{-3}\ R_n + 2.7x10^{-2}\ RH - 5.35x10^{-2}\ AW + 4.01x10^{10}\ (AW)^5 \qquad (7)$$

The statistical properties of this regression are shown in Table 1.

Table 1. Regression of ΔT on R_n, RH, and AW

Variable	Coefficient	Standard error	Students t value
Constant	-1.065	$5.02x10^{-1}$	- 2.12
R_n	$4.71x10^{-3}$	$3.69x10^{-4}$	12.75
RH	$2.74x10^{-2}$	$5.28x10^{-3}$	5.18
AW	$-5.35x10^{-2}$	$5.52x10^{-3}$	- 9.69
$(AW)^5$	$4.01x10^{-10}$	$6.18x10^{-11}$	6.49

Degrees of Freedom = 185
Residual Mean Square = .6754
Root Mean Square = .8219
R - Squared = .5902

Field Evaluation

Equation (7) was used to schedule irrigations on field corn during the 1980 growing season by setting AW equal to 50%, measuring R_n, RH and ΔT then comparing values of the measured ΔT with those calculated using Eq. (7) with measured values of R_n and RH. Irrigation was initiated when the measured ΔT value was greater than or equal to the calculated ΔT value. This scheduling procedure was compared to a checkbook method (Werner 1978), (soil moisture accounting and simple ET modeling), use of electrical resistence blocks to determine 50% depletion of AW and an unirrigated check. The experiment was a randomized complete block with four replications. Comparisons were made both of crop yield and water applied. The mean yield of the irrigated plots was ten t/ha with no significant differences among the three scheduling methods. However, with the ΔT method only 15.88 cm of water were applied while with the checkbook method 26.04 cm were applied and 19.43 cm were applied using the resistance block method. A complete description of the experiment and detailed results may be found in Geiser et al. (1980) and Geiser (1981).

Refinement of Scheduling Approach

An attempt was made to refine Eq. (7) by repeating the regression analysis, including the 1980 data and by inclusion of days after emergence (DAE) as a variable. This analysis indicated that (DAE) was not a significant variable and the form of the regression equation remained essentially the same as Eq. (7).

It is recognized that Eq. (7) is of little value at locations other than the experimental site, primarily because the effect of soil-water is expressed in terms of percent of available water. Since crop-water stress is more directly related to soil-water potential (Ψ_s) than to AW, a scheduling procedure utilizing Ψ_s rather than AW as the soil parameter should be applicable over a wider range of soil types. To develop a Ψ_s based regression equation, values of AW were converted to Ψ_s values using a weighted soil-water-characteristic curve. Weighting was necessary since the 91 cm depth of soil used as the irrigation depth was nonuniform. The top 22.5 cm of soil contained 41% of the AW between -1/3 bar and -15 bars while the bottom 25 cm of soil contained only 7% of the AW. The weighted relationship between AW and Ψ_s is shown in Table 2.

Table 2. Weighted Relationship Between Available Water and Soil-Water Potential for 91 cm Hubbard Sand Profile

Soil-water potential Ψ_s (bars)	Available water AW (%)
-0.33	100.0
-1.0	41.7
-3.0	15.2
-6.0	8.0
-9.0	5.5
-15.0	0

A regression analysis of the three years of data was performed with the absolute value of soil-water potential ($|\Psi_s|$) as the independent variable for soil-water. Other variables were as before. The following relationship is the result of this analysis:

$$\Delta T = 5.57 + 4.66x10^{-3}\ R_n + 4.19x10^{-2}\ RH + 3.85\ |\Psi_s| - 14.06\ |\Psi|^{0.2} \quad (8)$$

The statistical properties of this regression are shown in Table 3.

Table 3. Regression of ΔT on R_n, RH, and $|\Psi_s|$

Variable	Coefficient	Standard error	Students t value
Constant	5.57	2.26	2.47
R_n	$4.66x10^{-3}$	$3.28x10^{-4}$	14.23
RH	$4.19x10^{-2}$	$3.97x10^{-3}$	10.55
$\|\Psi_s\|$	3.85	$5.28x10^{-1}$	7.28
$\|\Psi\|^{0.2}$	-14.06	2.86	- 4.92

Degrees of Freedom = 245
Residual Mean Square = .8449
Root Mean Square = .9192
R - Squared = .5978

While application of Eq. (8) is still limited by the crop and climatic conditions under which it was developed, it should be possible to apply it to corn in climatic areas similar to central Minnesota over a wider range of soil types. Of course, it should be pointed out that the approach has not been tested on other soil types.

The use of the equation and the effect of R_n and RH on ΔT can be illustrated by setting $|\Psi_s|$ constant and solving for ΔT as a function of the other variables. Thus, if the desired level of soil-water depletion is 50% for the Hubbard sand, this corresponds to a $|\Psi_s|$ of 0.85 bars. Figure 1 is a graphical representation of the solution of Eq. (8) with $|\Psi_s|$ = .85 bars and RH as a parameter. From this plot it can be seen that at a level of net radiation of 500 $w \cdot m^{-2}$ and a relative humidity of 50%, the critical ΔT would be about -0.5°C (e.g., $T_c < T_a$) and at 600 $w \cdot m^{-2}$, it would be +0.25°C. In the first case, irrigation would be indicated if the value of ΔT measured with the infrared thermometer was -0.5°C or greater while in the second case the measured ΔT could be +0.25°C before irrigation would be required. In either case, the irrigation amount would be 50% of the available water or 3.45 cm in the case of the Hubbard sand. However, in a subhumid area where rainfall may be likely, it is desirable to fill the profile to something less than field capacity to allow for rainfall. In the experiments previously described, 2.54 cm of water was applied bringing profile available water to 87%.

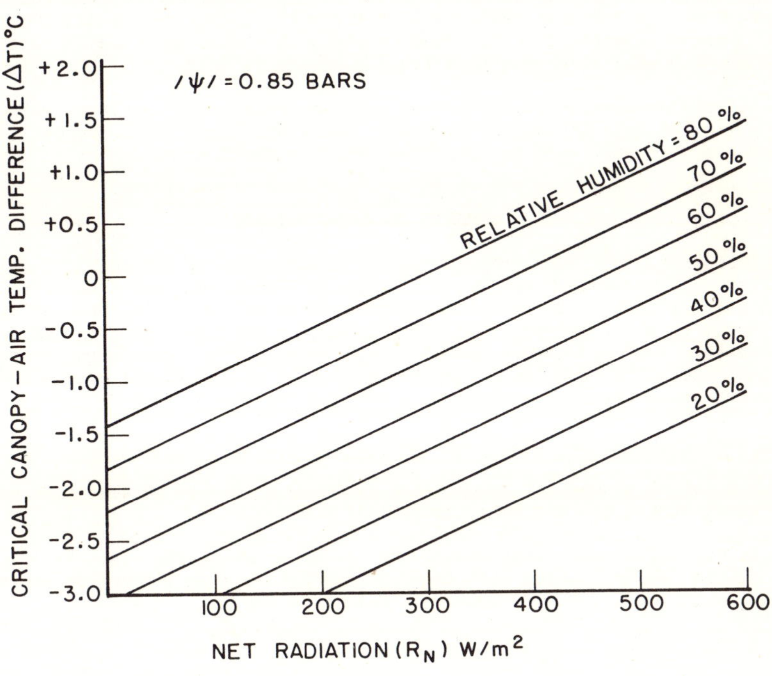

Fig. 1. Graphical Representation of $|\Psi_s|$ Based ΔT Model

Alternatively, a graphical representation similar to Fig. 1 could be constructed using any desired soil-water potential. Some level at a known soil-water depletion is convenient because it serves to indicate how much water may be added to the profile. Obviously, in a different soil 50% AW would most likely correspond to a soil-water potential different than the -0.85 bars used in the example.

Future Directions

A problem with putting the scheduling procedures described in this paper to practical use is that values of both R_n and RH are required in addition to the ΔT mesurement. Relative humidity measurements are relatively easy to make and should be no problem for a farmer to obtain. Net radiation appears to be more difficult, however, the current cost (1981) of the net radiometer used in this study is $700 and the infrared thermometer price is $800. Even if an irrigator were to purchase a small recorder for use with the net radiometer, he could obtain all of the equipment needed for scheduling for between $2500 and $3000 at 1981 prices. This cost is not unreasonable considering the probable investment to be made for the entire irrigation system.

An alternative to the regression approach described in this paper would be to use the threshold temperature approach. That is, irrigate when the ΔT increases above some threshold valve. However, Stange et al. (1981) and Jackson et al. (1981) indicate that this threshold is strongly influenced by net radiation and in regions such as the midwest where clear skies cannot be expected near solar noon each day of the growing season, net radiation may vary greatly from day to day. Thus, to be workable in this region, a threshold approach will also have to allow for variation in R_n.

The regression approach described in this paper is under continued refinement and a similar relationship is currently being developed for potatoes. The Ψ_s based approach will also be tested on finer textured soils.

SUMMARY AND CONCLUSIONS

The basis for use of the crop-canopy air temperature difference (ΔT) as an indicator of crop-water stress for irrigation scheduling was illustrated. An equation relating ΔT to other significant variables was developed and used as the basis of an irrigation scheduling procedure for field corn in central Minnesota. This approach compared favorably with other scheduling methods in current use. The equation was modified to make it applicable over a wider range of soils within the Midwest.

From this study it was concluded that:

(1) The use of canopy-air temperature difference (ΔT) as an indicator of crop-water stress has a strong theoretical basis.

(2) An irrigation scheduling procedure utilizing ΔT as the basis for the initiation of irrigation also needs to account for the effects of net radiation (R_n) and relative humidity (RH).

(3) A regression relationship expressing ΔT as a function of R_n, RH and available soil-water (AW) was used successfully to predict irrigation need and compared favorably with other scheduling methods.

(4) Continued refinement of the ΔT scheduling procedure should make it more widely applicable to a variety of crops and soils.

ACKNOWLEDGEMENTS

The research reported in this paper was funded in part by the Minnesota Agricultural Experiment Station Project MIN-12-45 and MIN-12-47. Additional support was provided by the Office of Water Research and Technology, U. S. Department of Interior.

REFERENCES

1. Anonymous, 1980. Irrigation survey. Irrigation Journal 30C67:72A-72H.

2. Barthekur, N. 1975. Leaf temperatures under controlled environmental conditions. J. Biometeor. 19:21-27.

3. Beadle, C. L., K. R. Stevenson, H. H. Neumann, G. W. Thurtell and K. M. King. 1973. Diffusive resistance, transpiration, and photosynthesis in single leaves of corn and sorghum in relation to leaf water potential. Can. J. Plant Sci. 53:537-544.

4. Blad, B. L., B. R. Gardner, K. L. Clawson, N. J. Rosenberg, D. G. Watts, R. E. Maurer, D. P. Garrity, D. G. Wilson and S. Steinmetz. 1980. Remotely sensed crop temperature for water resources management. Project Completion Report Nebraska Water Resources Research Institute. Lincoln, NE. 253 p.

5. Carlson, R. E., E. O. Yarger and R. H. Shaw. 1972. Environmental influences on the leaf temperatures of two soybean varieties grown under controlled irrigation. Agron. J. 64:224-229.

6. Clawson, K. L. and B. L. Blad. 1980. Irrigation scheduling using infrared thermometry. Paper presented at the 1980 Annual Meeting of the American Society of Agronomy, Detroit, MI. December 1980.

7. Decker, W. L. 1964. Total energy budget of the plant canopy and its relationship to evapotranspiration from corn. Mo. Agric. Expt. Sta. Res. Bulletin 854. Columbia, MO.

8. Erhler, W. L. and C. H. M. van Bavel. 1967. Sorghum foliar responses to changes in soil-water content. Agron. J. 59:243-246.

9. Gates, D. M. 1964. Leaf temperature and transpiration. Agron. J. 56:273-277.

10. Geiser, K. M., D. C. Slack, E. R. Allred and K. W. Stange. 1980. Irrigation scheduling using crop-canopy air temperature difference. ASAE Paper 80-2568. Presented at the 1980 Winter Meeting of the American Society of Agricultural Engineers. Chicago, IL. 20 p.

11. Geiser, K. M. 1981. Irrigation scheduling using crop-canopy air temperature difference. M.S. Thesis. University of Minnesota, St. Paul, MN.

12. Grannes, S. G., D. C. Slack, and E. R. Allred. 1978. Utilizing an infrared thermometer for air temperature measurements. ASAE Paper No. 78-2548.

13. Hiler, E. A., T. A. Howell, R. B. Lewis, and R. P. Boos. 1974. Irrigation timing by the stress-day index method. Trans. ASAE 17:393-398.

14. Idso, S. B., R. D. Jackson and R. J. Reginato. 1977. Remote sensing of crop yields. Sci. 196:19-25.

15. Jackson, R. D., R. J. Reginato and S. B. Idso. 1977. Wheat canopy temperature: A practical tool for evaluating water requirements. Water Resources Res. 13:651-656.

16. Jackson, R. D., S. B. Idso, and P. J. Pinter, Jr. 1981. Canopy temperature as a crop water stress indicator. Water Resources Res. 17(4):1133-1138.

17. Kanemasu, E. T. and C. B. Tanner. 1969. Stomatal diffusion resistance of snap beans. I. Influence of leaf-water potential. Plant Phy. 44:1547-1552.

18. Reicosky, D. C., D. E. Deton, and J. E. Parson. 1980. Canopy temperatures and evapotranspiration from irrigated and stressed soybeans. Agr. Meteor. 21(1):21-35.

19. Stange, K. W., D. C. Slack, E. R. Allred and K. M. Geiser. 1981. Relationship between canopy-air temperature difference and stomatal diffusion resistance. ASAE Paper 81-2010. Presented at the 1981 Summer Meeting of the ASAE. Orlando, FL.

20. Sumayao, C. R. 1979. Leaf temperature: an effective measurement in soil-plant-water research. Ph.D. Dissertation. Kansas State University. Manhattan, KS.

21. Turner, N. C. 1974. Stomatal behavior and water status of maize, sorghum, and tobacco under field conditions. Plant Phys. 53:360-365.

22. Werner, H. D. 1978. Irrigation scheduling checkbook method. University of Minnesota Agricultural Extension Service. Bulletin M-160.

AN AGRO-ENVIRONMENTAL MONITORING SYSTEM

S. D. Shaffer, N. L. Powell, and T. Martin

INTRODUCTION

Technology has provided modern agriculture with numerous resources - including chemicals for pest control and equipment for irrigation - which can be used to respond to the detrimental effects of variables in weather and disease population size. Since the use of these resources in turn impacts the quality of the environment and the cost of food production, the need to specify the variables is apparent. One solution, taking advantage of recent advances in computer technology, is a monitoring system consisting of an automated network of field data acquisition stations and a central control and dissemination unit which can deliver timely and accurate information. This concept was first fully realized in Purdue University's Ag Data Network (Biesslin, 1980). Another such system, the Agro-Environmental Monitoring System (AEMS), was developed under a NASA transfer in technology grant specifically to benefit Virginia's agricultural community. (Figure 1)

It is the intent of this paper to describe the AEMS system. After a brief overview, it will detail those features considered to have been most significant in contributing to the success of this network.

OVERVIEW

Information enters the network at six field stations located at key agricultural research centers. Data from a variety of sensors are collected under microcomputer control, converted to digital numbers, processed into time weighted averages, and stored at ten minute intervals. These data are then transmitted to the central control station in response to daily telephone interrogation. In addition to managing communications with the field stations, the central control station performs fundamental diagnostics, converts the data into common units of measure, archives the results, and generates primary output in the form of printed listings and graphic plots. The information is finally processed by applications programs including water balance and disease prediction models and made available to various users through the university's central computing facility.

FIELD STATION

The field station constitutes the front end of the information network. It has to accurately obtain data from a variety of signal sources, secure it in volatile memory for at least 24 hours, and then transmit it over telephone lines as much as 350 miles across the state of Virginia. Since the data are used in continuous simulation models to predict the stage of complex biological processes and environmental conditions, the station has to execute its functions reliably and without interruption, unattended and vulnerable to weather conditions. Hardware development for the field stations, therefore, demanded the most effort at the beginning of the

The authors are: S. D. Shaffer, Research Associate, N. L. Powell, Associate Professor, and T. Martin, Systems Analyst, Department of Agronomy, Virginia Polytechnic Institute and State University.

Figure 1. AGRO-ENVIRONMENTAL MONITORING SYSTEM

*Future Stations

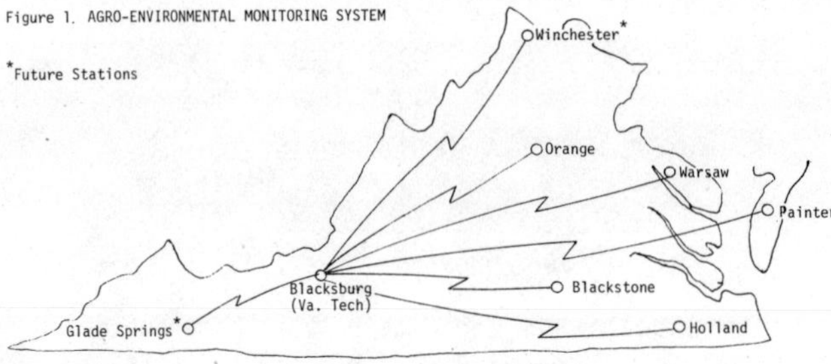

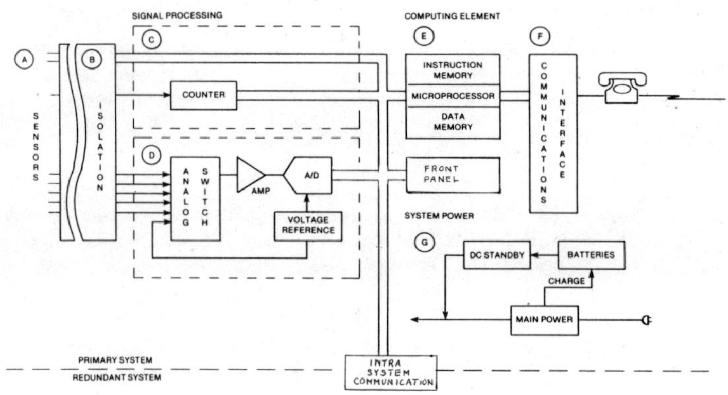

Figure 2. Field Station Block Diagram

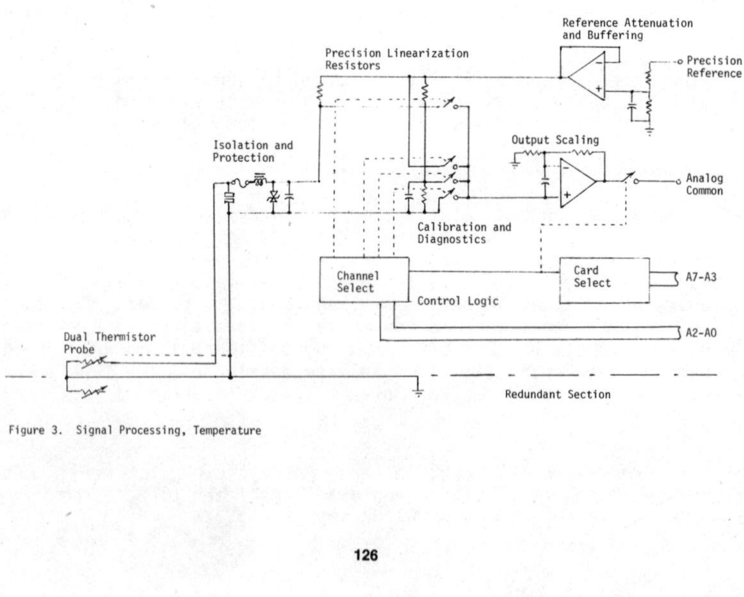

Figure 3. Signal Processing, Temperature

program. The component parts of each station include the following (Fig. 2):

- sensors to convert physical quantities to analog and digital signals (A)
- input isolation to filter the signals and protect electronic components (B)
- multiplexers or switch-arrays to select one signal at a time (D)
- amplifiers to convert analog signals to voltages of similar range
- an analog to digital converter to generate numeric quantities
- a microprocessor to control the step by step field operation (E)
- memory to hold the list of these steps and to store data
- a communications modem for telephone data transmission (F)
- a front panel for status display and diagnostics
- a power source with battery back-up (G)

Sensors

Environmental factors to be recorded, accuracy and durability of the sensor, availability at acceptable cost, and ease of interface with automated equipment were considerations given to sensor selection. The final decision included the following:

- wind speed - an optically coupled anemometer
- wind direction - a linear potentiometer
- barometric pressure - a piezoelectric load cell
- ambient temperature - a precision dual thermistor
- dew point temperature - a LiCl dew cell and dual thermistor
- soil temperature (at 2, 4, and 18 inches) a dual thermistor
- precipitation - a tipping bucket with reed relay
- solar radiation - a cosine corrected pyranometer
- photosynthetic active radiation - a photodiode with bandpass filter
- ozone - a photon counter to determine characteristic U. V. absorption
- leaf wetness - a Hughs-Davis (Davis, 1969) moisture sensing grid

The Hughs-Davis leaf wetness sensor is of particular interest to our research. Although its output, a log-log resistance, does not precisely relate to the amount of moisture on a leaf surface of a given plant, it can be used to determine accurately the time and duration of leaf wetness as well as detect trace precipitation that is insufficient to trip the rain gauge sensor. Efforts are now underway to define what influence this input might have on the current models that predict the emergence of pests and the need for irrigation.

Accuracy

A key field station objective in this design was to make signal processing and conversion elements sufficiently accurate so that the sensors would be the main source of error in the data stream. This was achieved to some extent by making many of the additional sources of error predictable and taking advantage of AEMS network computing elements to calculate these errors back out (Figure 3). It involved placing the multiplexer switch between the sensor outputs and the signal processing amplifier. A precision reference and ground potential were then provided as two additional sensor inputs so that they also would be affected by the same electronic pathway. Gain and offset errors would thus be collected along with sensor values and used to determine the absolute level of unknown signals. This ratiometric approach is identical to the laboratory situation in which one would measure an unknown quantity after carefully nulling the instrument to be used, and calibrating it against a high quality standard.

Environmental Protection

Despite the basic reliability of the equipment used, electronic components are still subject to several hazards such as extremes of temperature and humidity and electrical surges from lightning and other sources. Temperature inside the field station equipment shelter is maintained within a ten degree centigrade range by an air conditioner and space heater and humidity is minimized simply by component self heating.

The most disastrous threat is lightning. At its subtlist level a momentary or long term processing confusion sets in, the latter requiring a manual reset. The worst situation is the almost complete destruction of the station's electronic equipment. The problem of small surges is handled with software. To guard against the more intense surges, hardware is required.

Although the probability of a close hit is diminished by a ground charge dissipation array mounted on the wind tower, lightning induced surges can enter the equipement at three points: the power lines, the telephone lines, and through sensor cables. Power line transients are minimized with a high speed voltage clamp, an auto-resetting thermal relay and an equipment line filter. Protection from telephone line surges is provided by spark-gap devices that limit long (several millisecond) surges to 600 volts. This problem is further alleviated at two stations where the telephone coupler signals are treated in the same way as are the sensor signals.

The sensor lines initially relied on sub-nanosecond, 8 volt transient absorbers, overvoltage tolerant interface circuits, and electromagnetic shielding for protection. Later a single-point-ground electrostatic shield was added. Finally, in the summer of 1980, maximum protection from surges entering through the sensor cables was installed at two stations. It included a 90 volt spark-gap in series with a fuse that shorts the input side to ground when tripped and a 4 millihenry toroid in front of the transient absorber which is at the signal conditioning card edge for each input (Figure 3). Since that time, no surge related problems have occurred at these stations.

Power Back-Up

A power reserve system was developed to maintain continuity of operation in the field. A small AC-standby for devices requiring 110 volt AC is switched in by a relay when it is needed. But, to avoid relay switching transients, the computer, requiring only DC voltages, was backed up with a battery powered reserve that operates concurrently with the primary DC power supply. In order to keep the batteries charged and at the same time constantly available in the event of a power failure, they have been wired to the unregulated lines of the primary sources where they operate at a voltage slightly lower than the primary. Because there is no need for a switching element to change from one source to the other, transitions are smooth and virtually undetectable in either the digital or analog circuits. This approach is conceptually more straightforward and more reliable than the commercial uninterruptable power supply (UPS) since here the two sources are largely independent and operate in parallel with one another.

Station Back-Up

To further assure data continuity a back-up unit has been designed and built as a mirror image of the primary system. Both halves constitute the field station. They work in complete electrical isolation from each other except for an intra communications link (Figure 2).

At this writing only one of the halves is operating. Most of the other half has been installed at each station, but now provides only an on-site, card-level spare parts inventory. In the near future, when the back-up half is operational, it will serve as a check of the data already on line as well as an alternate source of that data. Finally, when intra communication is implemented, error detection will be enhanced with local diagnostics, and malfunctions will be better survived by the ability of each half to share resources with the other.

Communications

Custom field communications equipment was designed for this project in an effort to streamline the equipment inventory. It was encouraged by the availability of numerous subsystem components that would simplify this effort. The final product was a single, low-power board that interfaced directly to the computer bus at one end and to the telephone coupler at the other. A pulse dialer to enable the field to originate calls in an emergency and a RS-232 port to facilitate maintenance were also included on the board. This device is complemented at the other end of the telephone link, the central control station, with a commercial, 300 baud, Bell 103 modem.

CENTRAL CONTROL STATION HARDWARE

A number of minicomputer suppliers offered ready made equipment from a central control station which could be configured. Therefore, no hardware design effort was required at this point. The final selection of components was predicated on the biases and experience of our personnel as much as our need for facile software development and a high volume data management. The current configuration includes:

64 Kbyte, 16 bit minicomputer
32 bit precision floating point hardware
50 Mbyte, removable pack disc drive
CRT terminal and back-up hard copy terminal
9-track magnetic tape drive
300 LPM printer
CRT terminal and digital plotter for color graphics

SOFTWARE

The software has the task to supervise field station activity, to direct the flow of information in the network, and to determine the form of data at each processing node. The AEMS software makes up three groups, the: field station software, the central control station software, and the CMS-IBM software.

The field station software functions were chosen to be as simple as possible in order to minimize the influence of unattended field equipment on the data gathered, and to reduce the assembly language programming effort and the amount of memory required. The final field package is responsible for the execution of four key operations:

initial capture of transducer and reference signals
maintenance of time weighted averages of each signal
filing of these averages with a time stamp at ten minute intervals
transmission of this file to the central control point, located in the Agronomy Department at Virginia Tech, when requested

It was decided that hardware interrupts and DMA (direct memory access) features would be disabled and that all software functions be carried out in a round-robbin or polling scheme. In this manner reliance is placed mainly on the processor and its memory, and continuity of operation is relatively immune to peripheral hardware malfunctions.

The central control station software is responsible for:

communication with the field stations as well as with the university's IBM computer
conversion of raw sensor data into common units of measure
generation of listings, graphs, summaries and model output
updating of the data base.

Communication between the field and the central control site required the coordinated development of software routines to operate each end of the link. Advantage is taken of the fact that the data flow is primarily in one direction and a straightforward but effective protocol is implemented. Vertical and horizontal redundancy codes are generated at the field station for each line of data and transmitted with the data line. They are checked by the central unit to detect transmission errors. The central station responds with a brief acceptance message requesting either new data or, in case of an error, retransmission of the previous lines. Communication between the central control station and the IBM computer is established through Interdata's Telecommunications Access Method (ITAM) on a dial up 2400 baud synchroneous line.

This link enables the maintenance of the AEMS data base also on Virginia Tech's main computer. Once stored on the main computer, the information is then accessible to anyone, with a legitimate use for the data, through a software package. This package is available under CMS - IBM's interactive system. The number of programs to choose from is growing depending on the request for different data groups needed. At this writing programs include:

output of daily data from all sensors in ten minute intervals
output of daily data from all sensors averaged over two hours, max-min-average daily temperature, average daily relative humidity, average daily dew point temperature, total daily precipitation, total daily soar radiation and total daily photosynthetic active radiation
daily max-min of the ambient temperature and soil temperatures at 2", 4", and 18" with their time of occurance
duration of the ambient or soil temperatures below and above certain thresholds.

The software for the field and central control station is developed on the Interdata computer. It has a well functioning operating system, editor, assembler, FORTRAN compiler, and cross assembler which generates the machine code of our field station. The interactive package is developed on the IBM using FORTRAN and PL/I.

CONCLUSIONS

The value of the AEMS information has been realized particularly in the research of pest management and irrigation scheduling models. Continuous, accurate, and high resolution data is now available from a broad range of relevant parameters at the specific locations where these research activities are carried out. The results of this research were sufficiently encouraging so that a leaf spot spray adivsory for peanuts was distributed through radio and television media during the summer of 1981 and an irrigation advisory is planned for 1982 (Powell, 1981). Currently, two other advisories, one for planting and the other for the harvesting of peanuts, are also available to Virginia's growers.

The risk of extrapolating too far from this site specific data has been somewhat reduced by the correlation and verification obtained from volunteer weather observers throughout Virginia who keep track of temperatures, rainfall, and cloud cover conditions. In the future this danger will be further reduced with the incorporation of hourly visible and infrared satellite imagery. This will not only provide radiation and temperature estimates but also, due to recent efforts at NASA, it will give us rainfall estimation as well. All of the algorithms have been developed and verified and the programming is well underway.

REFERENCES

D. R. Davis and J. E. Hughes, "A New Approach to an Old Technique for Measuring and Recording the Wetting Parameter in Vegetable Covers", manuscript of the ESA Weather Bureau, 1969.

N. L. Powell, B. B. Ross, F. S. Wright, G. D. Jones and V. O. Shanholtz, "Evaluation of an Irrigation Scheduling System for Virginia" ASAE conference proceedings, 1981.

H. Biesslin, "Project Background FACTS and Extension", proceedings FACTS Computer System and Update Conference, September 16 and 17, 1980.

IRRIGATION SCHEDULING IN HUMID AREAS

J. R. Lambert, C. W. Doty, and V. L. Quisenberry
Member ASAE Member ASAE

Irrigation of humid areas of the U.S. was very sporadic until the middle to late seventies. Farmers irrigated only when moderate to severe drought occurred except for horticultural or specialty crops. The need for reliability of crop yield, cash flow income to insure annual net profits, and increased emphasis on national production has increased the emphasis on southeastern agriculture, and renewed interest in irrigation. For example, total irrigated acreage in Georgia has increased from 58,500 ha in 1970 to an estimated 340,000 in 1979.

Climatic variability is typified by nonirrigated corn yields which varied from 0 kg/ha in 1954 to 8,230 kg/ha (131 bu/ac) in 1950 in an irrigation experiment in South Carolina. Yield was 2,000 kg/ha (32 bu/ac) or below for 6 years and 6,000 kg/ha (96 bu/ac) or above for 6 years out of the 19 years. Such variability plays havoc with a farmer's cash flow, and usually with the overall profitability of his farm. As the cost-price squeeze has tightened during the last few years, and as low-labor requiring systems have become available, many farmers in the humid areas of this country have installed irrigation systems. This has been particularly true in the southern areas where plentiful water, abundant radiation, longer growing seasons, and higher temperatures are conducive to increased production of a wide variety of crops.

Irrigation in humid areas is often economical even though annual rainfall exceeds evapotranspiration. Three factors necessitate irrigation of humid areas: (1) the annual rainfall distribution does not coincide with the evapotranspiration distribution, (2) water holding capacity generally is not sufficient to provide adequate water for crops during the deficit rainfall periods and (3) frequently restricted rooting limits soil water availability to plants.

Average monthly rainfall and evaporation at 4 representative locations in the eastern U.S. are shown in Fig. 1. These locations typify the widespread humid areas where deficit rainfall occurs throughout the entire growing season. Rainfall is much more erratic than evaporation. Less than 5 cm of rain may fall during any month of the growing season (Fig. 1). Evaporation rates near 15 cm/mo cause very serious deficits.

Most soils in the southeastern U.S. are of relatively low water-holding capacity and the crop root systems are often shallow due to physical impedance, chemical toxicity or poor aeration. Little water is held at water potentials less than -1 bar (Bruce et al. 1980). Campbell et al. (1974) found that the Ap horizon, or surface layer, of typical Paleudults of the South Carolina

---

The authors are: J. R. Lambert, Professor, Agricultural Engineering Department, Clemson University, Clemson, SC; C. W. Doty, Agricultural Engineer, Coastal Plains Soil and Water Conservation Research Center, USDA-SEA-AR, Florence, SC; and V. L. Quisenberry, Associate Professor of Agronomy, Clemson University. Technical Contribution No. 1978 of the South Carolina Agricultural Experiment Station, Clemson University.

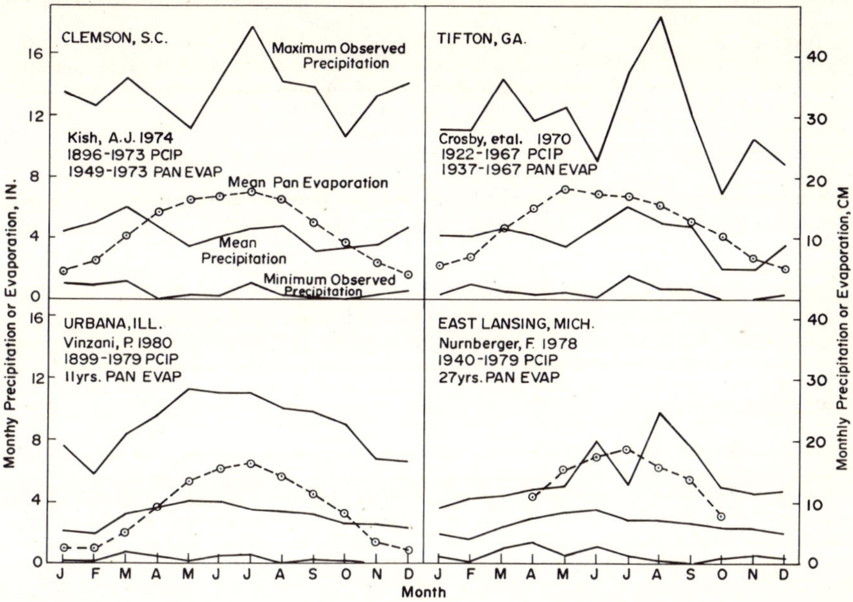

Fig. 1. Monthly Rainfall Distribution and Average Pan Evaporation for 4 Humid Locations.

Coastal Plains drains 42% of its pore volume between -0.1 and -1.0 bars. Soil water retention between -0.05 and -1.0 bar ranges from 0.05 to 0.13 cm/cm for both Piedmont and Coastal Plain soils (Bruce et al. 1980). For 70-cm rooting depth, 9.1 cm is the maximum water holding capacity available in most of these soils. Typically used values are 5 to 8 cm. With peak evapotranspiration rates of 0.6 to 0.8 cm/day usual irrigation intervals are 5 to 7 days.

The purpose of this paper is to describe four methods now being tested for scheduling irrigation and to discuss experiences related to their use. This study has been partially funded by USDA-AR at Florence, South Carolina, and influenced by previous work by Jensen et al. in the West. Modifications and additional studies have been necessary because of differences in water-holding capacities of the soils, lack of validation of evapotranspiration models, and interspersed rainfall.

SCHEDULING VIA PERSONAL COMPUTER AND WATER BUDGET

Increased irrigation, especially by inexperienced persons, has led to renewed interest in scheduling methods. Low-cost personal computers that are available to individual farmers, extension agents, and consultants have influenced a study to determine the feasibility of using these computers and the water budget approach on individual fields to reduce water and energy consumption, improve economic yields, and develop a tool that individual farmers or consultants can use. Such a practice must be kept relatively simple. The philosophy of this study is that the practitioner will run the program on his own computer using data he acquired by observing his fields or from the National Weather Service information.

Approach

Water budgets are calculated semi-weekly using historical data for the previous 3 or 4 days on daily maximum and minimum temperatures, incoming solar radiation, effective rainfall or irrigation applied, and root-zone depth. The practitioner also states daily the allowable depletion as a percentage of total available soil water, based on crop growth stage, experience, and any other available subjective input. Initialization includes a one-dimensional description of soil horizon depths, water holding capacities, and initial water contents.

Daily evapotranspiration is calculated by the Jensen and Haise (1963) method and modified for canopy cover and soil moisture content. A water budget is maintained for the root zone, and reported in both inches and percent (Fig. 2).

IRRIGATION SCHEDULING FOR EDISS

07/31/80 11:45:15

BACK RECORDS:

DATE	TMAX	TMIN	RAIN	IRRG	WIND	RAD	PEVP	ETP	AETP	RD	AD(%)	AWC(%)	AWC(IN)	DPL(IN)
07/28	94	68			19	379	.216	.201	.178	42.0	50	72.0	3.64	1.42
07/29	94	68			24	391	.215	.207	.179	42.0	50	68.4	3.46	1.60
07/30	95	68			24	403	.215	.215	.183	42.0	50	64.8	3.28	1.78

FORECAST:

DATE	TMAX	TMIN	RAD	ETP	AETP	RD	AWC(%)	AWC(IN)	DPL(IN)	
07/31	91	67	391	.201	.167	42.0	61.5	3.11	1.95	
08/01	94	68	437	.231	.188	42.0	57.8	2.92	2.14	
08/02	95	70	415	.225	.178	42.0	54.3	2.75	2.31	
08/03	97	73	440	.247	.190	42.0	50.5	2.56	2.50	
08/04	92	72	400	.215	.161	42.0	47.3	2.40	2.66	IRRIGATION NEEDED.

Fig. 2. Typical Output from Water Budget Program for Personal Computer Used for Irrigation Scheduling.

A Radio Shack* TRS-80/I computer was programmed to interactively request the user to input daily TMAX and TMIN (deg F), any effective RAIN or IRRiGation (in), WIND (mi), incoming solar RADiation (ly), observed Rooting Depth (in), and Allowable Depletion (%) from the date of previous analysis (28 July) until "yesterday" (30 July), the last day for which data are available. The program is very user oriented, with error trapping, safeguards, and explanatory comments. From the input data, Potential daily EVaPotranspiration (in), EvapoTransPiration (in) considering canopy cover, and Actual EvapoTransPiration (in) considering both canopy cover and soil moisture Content, Available Water Content (% and in) and DePLetion (in) are calculated. If the analysis indicates that the depletion exceeded the stated allowable depletion on a previous day, the corresponding line is flagged during output. A field named EDISS had a depletion of 4.52 cm (1.78 in, 35.2%) in a 107-cm (42-in) root zone on 30 July 1980 (Fig. 2).

Similar calculations are made for a future 5-day period, using quantitative 5-day forecasts of maximum and minimum temperatures and radiation furnished for six locations throughout the Southeast by the Columbia, South Carolina

* Mention of trademark, proprietary product, or vendor does not constitute a guarantee or warranty of the product by the U.S. Dept. of Agric. or the S.C. Agri. Exp. Sta. and does not imply its approval to the exclusion of other products or vendors that may also be suitable.

National Weather Service. Rainfall is precluded from the 5-day projected budget. The resulting budget predicts depletion for the 5-day future assuming no rain. The first day during which depletion is forecast to exceed the stated allowable depletion is flagged to indicate that irrigation will be needed near that date (4 August). If no irrigation need during the next five days is forecast, a straight-line extrapolation of depletion is made and the date of anticipated irrigation need is stated on the printout.

If rain does fall during the forecast period, the budget may be adjusted manually or mentally, or the program may be rerun to determine if any forecasted irrigation is needed. Reinitialization of water content profiles may be done at any time data are available.

Experience

Cooperating researchers at five locations in four southeastern states have irrigated corn and soybeans during 1979, 1980 and 1981 by the personal computer/water-budget approach. Budgets were calculated on Monday and Thursday mornings, typically, based on temperatures, radiation, and rainfall observed at the field site.

Mechanical and quantitative operation of the hardware and approach have been satisfactory. Comparison of predicted and measured soil water content data indicates the real need to periodically obtain field data for reinitialization. Whenever 3- to 4-week reinitializations are made, and whenever field scheduling of irrigations follows the analysis results closely, agreement of field-measured and computer-based soil water content has been acceptable.

Some underestimates of water needs were apparent, especially on sandier soils, but in most cases needs were adequately met. We obviously don't know enough about how to estimate available water in a profile; how to account for dry conditions in the upper, more densely rooted portions of the root zone while the lower portions are still wet; how to calculate evapotranspiration, especially under limited soil water contents; or how to manage the allowable depletion parameter to optimize crop behavior.

Results

Trials during a very dry 1980 at five locations resulted in 6 to 12 irrigations of 11.4 to 36.9 cm total applied water. Corn yields ranged from approximately 1,380 to 4,760 kg/ha (22 to 76 bu/ac) for no irrigation and 4,140 to 8,410 kg/ha (66 to 134 bu/ac) for irrigation by the water budget method. Field results from this project have been limited to areas that can be irrigated in one day.

SCHEDULING VIA SCREENED PAN EVAPORATION

The evaporation pan provides a way to physically simulate a water balance in the soil profile and schedule irrigations, using a specially equipped evaporation pan. Campbell and Phene (1976) showed that in the Southeast the amount of evaporation from a screened standard Class A pan is equal to Potential EvapoTranspiration (PET) as calculated by the Penman (1948) equation. By using screened pan evaporaton as PET, and a crop coefficient, Doty (1980) showed that the storage in the soil profile could be closely approximated. Based on these findings, the screened evaporation pan was modified to schedule irrigations and tell the farmer how much water to apply and when to irrigate. The assumption must be made when using the screened evaporation pan to schedule irrigation that water evaporates from the screened evaporation pan at the same rate as PET.

How Much to Apply Each Irrigation

To determine the amount of irrigation water to apply at each irrigation the following equation is used:

$$I = (A \times La) / E \qquad (1)$$

I = Depth of irrigation water to apply
A = Available water in the rooting depth (determined from SCS, National Cooperative Soil Survey, Blue Sheets of Soil Series description and Crop rooting depth)
La = The percent of available water to be used by evapotranspiration before irrigation is needed (Loss Allowable)
E = Efficiency of irrigation system.

When to Irrigate

The evaporation pan automatically indicates when to irrigate. The pan is modified with an overflow to discharge excess water analogous to runoff or deep percolation when the soil profile becomes full (Fig. 3). A movable stainless steel scale is also added. The amount of pan evaporation that must occur before irrigation is needed can be determined by the following equation:

$$Ad = (A \times La) / C \qquad (2)$$

Ad = Pan evaporation to occur before irrigation is needed (allowable depletion)
C = Ratio of actual evapotranspiration to screened pan evapora evaporation. This is known as crop coefficient. Normally we use C = 0.6 until the corn is above knee high. After that, we use C = 1.0. However, C can be changed more often if necessary.

Figure 3. Evaporation Pan Modified with a Screen, an Overflow and Scale to Schedule Irrigation for a Center Pivot System at Florence, SC.

To avoid estimating the amount of water stored in the profile when the schedule is started, the soil profile should be full. This can be accomplished by (1) setting up the evaporation pan the day after a rain that fills the soil profile, or (2) filling the soil profile by irrigating at the beginning of the irrigation season. The pan is leveled and filled to overflowing. A scale is then placed on the side of the pan and inserted into the water the depth of the allowable depletion (Ad in Eq. 2) and clamped to the side of the evaporation pan. The screened pan is observed as required, and when the water level of the evaporation pan drops to the end of the scale, the allowable water has been depleted by evapotranspiration and I depth of water (Eq. 1) must be applied by irrigation.

The evaporation pan will automatically adjust to the water balance throughout the season. If rainfall occurs, the pan catches it and excess water is removed from the pan by the overflow. However, if the evaporation pan is not placed under the irrigation system, the pan must be filled to the overflow point after each irrigation.

This scheduling technique requires that a separate pan be used for each crop being grown. The technique is not exact but is as accurate as most systems are able to apply water. Since only a portion of the available water is allowed to be depleted, this scheduling technique should call for water before the crop suffers.

Experience and Results

The screened evaporation pan was used to control a center pivot system for three years. In 1978 irrigation was scheduled by the screened evaporation pan for early and late soybeans. Even though at least one irrigation was missed because of malfunction of the irrigation equipment, five irrigations resulted in 10.3 cm of irrigation water being applied. Rainfall was 21.2 cm. Nonirrigated soybeans produced 1,660 kg/ha (26 bu/ac) for the early-planted soybeans and 1,190 kg/ha (19 bu/ac) for the late-planted soybeans, while the irrigated soybeans produced 2,430 kg/ha (39 b/ac) for the early-planted and 1,820 kg/ha (29 bu/ac) for the late-planted beans.

In 1979 (a near average rainfall year) corn required 5 irrigations, totaling 19.4 cm. Rainfall amounted to 47 cm during the corn growing season and 73 cm during the soybean growing season. The soybeans were irrigated 5 times with 19.8 cm of irrigation applied. Corn yields of 6,460 kg/ha (103 bu/ac) and soybean yields of 1,400 kg/ha (21 bu/ac) were harvested from the nonirrigated area. Irrigated yields under the center pivot system with applications scheduled by the screened evaporation pan were 10,900 kg/ha (174 bu/ac) corn and 2,450 kg/ha (36 bu/ac) soybeans. In 1980, a dry year, 29.7 cm of rainfall on corn and 43.3 cm on soybeans resulted in 6 irrigations on corn, totaling 25.6 cm; 12 irrigations on soybeans, totaling 34.9 cm of irrigation. Nonirrigated yields were 2,990 kg/ha (48 bu/ac) corn and 1,380 kg/ha (21 bu/ac) soybeans, while the evaporation pan scheduled irrigated area produced 6,050 kg/ha (96 bu/ac) corn and 2,400 kg/ha (36 bu/ac) soybeans.

SCHEDULING VIA TENSIOMETERS

For either the computer-based water-budget or the pan method to be used accurately, selected parameters must be specified: available soil water as a function of depth, rooting depth, and evapotranspiration rates throughout the growing season. However, in soils with similarly shaped soil water characteristic curves, a simple measure of the soil water tension will indicate the soil water status, eliminating the need for any other input.

Figure 4 shows desorption curves for three soil series found in the Piedmont and Coastal Plains of the Southeast. While the absolute values of the soil water contents differ greatly for given tensions, the general shapes of the curves are similar. The upper limit of available water in these soils is generally between -0.04 and -0.06 bars potential regardless of the horizon or texture. Little available soil water is held between potentials of -0.2 and -1.0 bars. Thus most of the available soil water is held between -0.05 and -0.2 bars. The only other information needed to use tensiometers to schedule irrigation is the soil depth over which the soil water potential is to be maintained at values greater than -0.2 bars. In this study, the soil water potential was kept above -0.2 bars throughout the 60 cm soil depth.

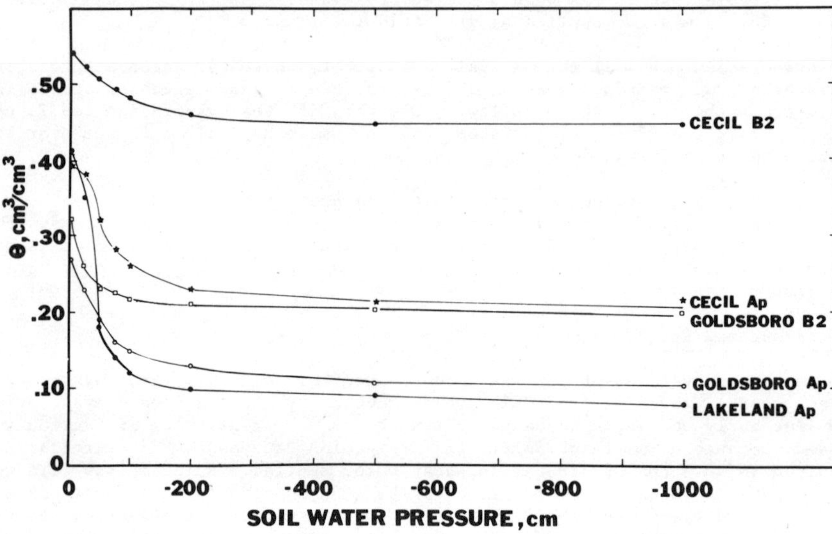

Fig. 4. Soil Moisture Characteristic Curves for Three Typical Southeastern Soils.

Experience and Results

During the past two years, experiments have been conducted at Blackville, South Carolina on a Wagram sand to compare the use of the computer-based water-budget method to tensiometers for scheduling irrigation of corn. A summary of the water use data for a period from June 10 to July 15, 1980, is presented in Table 1. Evapotranspiration was measured from tensiometers which were placed in 15-cm increments through 121.9 cm. During this 35-day period average daily ET was 0.33, 0.47 and 0.67 cm/day for the nonirrigated treatments, the water-budget treatment, and the -0.2 bar treatment, respectively. The water-budget consistently called for water 3 to 4 days after the need for irrigation was indicated from tensiometer readings. Often the water-budget corn showed severe stress symptoms before irrigation. With this type of information the water-budget and screened pan evaporation methods can be improved to meet the actual water needs.

There is no doubt but that some of the soil physical properties used in the water-budget were incorrectly used. In particular, a rooting zone depth of 137 cm was used throughout the period with an allowable depletion of 50%. There is also no doubt that the pyranometer has given low radiation readings, perhaps by 35 percent. Root samples taken frequently during this period showed rooting densities of approximately 0.4 cm/cm at the lower depth. Soil water potentials frequently decreased below -0.6 bars in the surface 45 cm before 50% of the available water through 137 cm had been used.

Table 1. Water Use Data from June 15 to July 10, 1980, for Corn Grown on a Wagram Sand.

Method	Rainfall	Irr.	Total	OO	Ave. ET	Ave. OP	% OP
	------------- cm		---------------		-------- cm/day ----		
0.2 bar	8.97	14.61	23.58	0	0.67	0.66	102
Budget	8.97	7.49	16.46	0	0.47	0.66	71
No-Irr	8.97		8.97	2.54	0.33	0.66	50

Another question regarding the water budget concerns the maximum ET estimated. Soil water use rates for corn during late vegetative and reproductive stages show that ET is approximately equal to open pan evaporation when the soil water potentials are maintained above -0.2 bars. As shown in Table 1, corn in the -0.2 bar treatment used approximately 100% of the amount of open-pan evaporation during this period. The water-budget treatment utilized 71% of open pan evaporation and the nonirrigated treatment used 50% of open-pan evaporation. Yield data show that the water-budget treatment was periodically stressed. Yields for the -0.2 bar, water-budget and nonirrigated treatments were 11,180, 6,670 and 4,780 kg/ha (166, 99 and 71 bu/ac), respectively. Based on these results soil water potentials must be maintained above -0.2 bars in the surface 60 cm for optimum corn growth to occur. ET must be maintained at about 90% of open pan evaporation during late vegetative and reproductive stages for maximum yields. Campbell and Phene (1976) showed that evaporation from the screened evaporation pan was about 90% of open pan.

Tensiometers provide an easy way to properly maintain soil water potentials within the desired limits for optimum crop growth. This is especially true for a crop such as corn for which just a few stressed days can result in substantially reduced yields.

COMPARISON OF THREE METHODS

Several plots under a center-pivot irrigation system at Florence, SC were controlled by three methods: water-budget, screened evaporation pan, and tensiometers. Table 2 shows the water applied and the yield for each scheduling technique. The screened pan required 2 applications (7.1 cm) more water than the water budget and 1 application (5.9 cm) more than the tensiometers in 1979. Corn yields in 1979 were 10,900, 10,170, and 8,370 kg/ha (174, 162 and 133 bu/ac) for the screened pan, tensiometer, and water-udget methods, respectively. However, factors such as poor estimation of rooting depth and allowable depletion and poor communications were problems with use of the water budget in 1979. In 1980 the water-budget and the tensiometer methods required 9 applications totaling 29.4 and 29.0 cm, respectively, while the screened pan method required 8 applications totaling 25.6 cm of water. Yields were 6,050, 7,730 and 7,770 kg/ha (96, 123 and 124 bu/ac) for the pan evaporation, water-budget and tensiometer methods, respectively. The pan evaporation method of scheduling irrigation increased yield compared to nonirrigated yields by 69 and 103 percent in 1979 and 1980, respectively; the water budget method increased yields by 30 and 159 percent; and the tensiometer method increased yields by 57 and 160 percent over nonirrigated corn.

The differences in water applied in 1980 (3.8 cm), the fact that there was only one application difference and the fact that the two-year average yield was similar (Table 2) indicate that all three methods of scheduling irriga- tion are feasible in the Southeast. However, there are several relations that still must be determined for more efficient irrigation scheduling by the water-budget and screened-pan evaporation methods. The relation of rooting depth to days after planting by soil type, system irrigation efficiencies, allowable depletions of soil water, and crop coefficients would assist these scheduling techniques to provide crop water needs, conserve water and energy, and increase irrigation production. When these are developed for each crop, the farmer can choose the method that best suits his needs for scheduling irrigation in the Southeast.

Table 2. Results from Three Irrigation Scheduling Methods on Corn that Was Under-row Subsoiled at Florence, SC

Year	Irrigation No. of Applications	Irrigation Yearly Amount	Rainfall	Total Water	Yield 1/	
		cm	cm	cm	kg/ha	% increase over nonirrig.
			Personalized Computer Model			
1979	3	12.1	47.4	59.5	8,365b	29.5
1980	9	29.4	29.7	59.1	7,730d	158.9
Mean	6	20.7	38.5	59.3	8,049	70.4
			Screened Pan Evaporation Method			
1979	5	19.2	47.4	66.6	10,895a	68.6
1980	8	25.6	29.7	55.3	6,048e	102.5
Mean	6.5	22.4	38.5	60.9	8,472	79.3
			Tensiometer Method			
1979	4	13.3	47.4	60.7	10,167a	57.3
1980	9	29.0	29.7	58.7	7,767d	160.1
Mean	6.5	21.1	38.5	59.7	8,967	89.8
			Nonirrigated			
1979	-	-	47.4	47.4	6,463c	-
1980	-	-	29.7	29.7	2,986f	-
Mean			38.5	38.5	4,724	

1/ Yields with the same letter within the same year are not significantly different at the 5% level.

SCHEDULING VIA CALCULATED RISK

The impetus for any farmer to irrigate, in a humid or non-humid area, is to increase the net return from a crop. Guessing, feeling the soil, measuring the soil water potential, and estimating the soil water budget are indirect approaches to decision making relative to the economic benefit of a particular irrigation.

Employing the principle of calculated risk (Thompson 1963) to make irrigation decisions involves calculation of an expected loss and comparing it with the cost that would be necessary to prevent the loss. The expected loss may be calculated by taking the product of the loss due to moisture deficiency that would be incurred, should no rain occur and no irrigation be applied, and the probability that no rain will fall. The cost of preventing the expected loss is the cost of the irrigation (Allen and Lambert 1971a).

The actual criterion involved, based on the calculated risk principle and applied to irrigation, may be stated as:

$$P \begin{array}{c} > \\ = \\ < \end{array} C/L \qquad \begin{array}{l} \text{irrigate} \\ \text{either course} \\ \text{do not irrigate} \end{array}$$

where P = probability of the loss occurring (no rainfall); C = cost of protective measures (irrigation) required to prevent the would-be loss; L = loss that would be incurred should no precipitation occur and no protective measures be taken. The loss (L) would be the decrease from potential in the final dollar value of the crop caused by non-optimum moisture conditions during a prescribed loss period subsequent to the decision. The attainable potential will vary throughout the season because of the residual effect of moisture during the earlier portion of the season and all other factors affecting final yield. Dynamic simulation models of crop growth and yield are used to estimate final crop yield under irrigation or no-irrigation scenarios (Lambert 1978).

Application of the calculated risk principle to daily decisions of irrigation has been shown to reduce the number of irrigations required, to reduce the total water added, and thus the total energy required, and to reduce the occurrence of 13 mm of rainfall within 24 hours after an irrigation, compared to a criterion of irrigation at 50% depletion of available soil moisture. Net economic return from tobacco also was improved (Allen and Lambert 1971b).

Field testing on corn has indicated the calculated-risk/crop-simulation method of scheduling irrigation in humid areas to have potential, especially now that marginal cost of irrigation has decreased significantly with the advent of automatically moving systems. Results of 4 years of experience with tobacco and corn indicate savings of water and energy compared to use of the 50% criterion for irrigation.

During the 1981 growing season we used the calculated risk method to schedule irrigation of corn at Tifton, Georgia. The program was a Fortran program written for the IBM 370, modified to run on the TRS-80/I.

SUMMARY

Poor rainfall distribution with respect to evapotranspiration distribution and low water-holding capacities of most soils cause irrigation to be needed in many humid areas, particularly in intensive agriculture. Rainfall occurs sporadically during the growing season and disrupts any preplanned schedule.

A consortium of researchers is evaluating techniques for scheduling irrigation in the Southeast. A personal computer has been programmed to calculate water budgets from data supplied by the user and to forecast the date of the next irrigation. The results are no better than the data supplied and evapotranspiration rates for southeastern U.S. conditions are not well known.

Addition of wire screen and an overflow to a Class A evaporation pan causes the evaporation to approximate the evapotranspiration from a nonstressed crop with full canopy cover. Such a pan can therefore be used to physically simulate the water budget of a profile and to indicate the need for irrigation when a preset depletion has been reached.

Tensiometers may also be used to determine the need for irrigation. Data indicate that soil water potentials maintained above -0.2 bar in the upper 60 cm produced maximum yields in these studies. Net economics of irrigation application is the basis for using the calculated-risk method for scheduling irrigation, including forecast probabilities of rain.

Based on initial tests, the pan method is the simplest and easiest to use, but is limited by the assumption of fixed rooting depth and gives little advance information for scheduling or planning purposes. The computer-based water-budget method is powerful for planning, but relies on soil physical parameters and evapotranspiration rates which we don't know how to determine well. Tensiometer methods apparently result in increased yields for the conditions tested, but require considerable attention and a decision on how many to use and where to locate them in the field and give little advance information. Inclusion of direct economic benefits and rainfall probabilities for scheduling irrigation is fundamentally more sound, but relies heavily on dynamic crop growth simulators which are still being developed.

REFERENCES

Allen, W. H. and J. R. Lambert. 1971a. Application of the principle of calculated risk to scheduling of supplemental irrigation. I. Concepts. Agr. Meteorol. 8:193-201.

Allen, W. H. and J. R. Lambert. 1971b. Application of the principle of calculated risk to scheduling of supplemental irrigation. II. Use on flue-cured tobacco. Agr. Meteorol. 8:325-340.

Bruce, R. R., J. L. Chesness, T. C. Keisling, J. E. Pallas, Jr., D. A. Smittle, J. R. Stansell, and A. W. Thomas. 1980. Irrigation of Crops in the Southeastern United States. USDA-SEA. ARM-S-9.

Campbell, R. B., D. C. Reicosky and C. W. Doty. 1974. Physical properties and tillage of Paleudults in the Southeastern Coastal Plains. J. Soil and Water Conser. 29:220-224.

Campbell, R. B. and C. J. Phene. 1976. Estimating potential evapotranspiration from screened pan evaporation. Agric. Meteor. 16:343-352.

Crosby, F. L., H. S. Carter, B. H. Quattlebaum, Jr., and Sam Burgess. 1970. Weather data analyses of the University of Georgia College of Agriculture Experiment Stations. Res. Rep. 66, Ga. Agr. Exp. Sta., Athens.

Doty, C. W. 1980. Crop water supplied by controlled and reversible drainage. Trans. ASAE 23(5):1122-1130.

Jensen, M. E. and H. R. Haise. 1963. Estimating evapotranspiration from solar radiation. Proc. ASCE IR 4:15-41.

Kish, A. J. 1974. Climate of Clemson. S.C. Agr. Weather Res. Series No. 43, Clemson Univeristy.

Lambert, J. R. 1978. Use of simulation in crop management: Irrigation. Proc. BSSG Workshop on Crop Simulation, Clemson.

Nurnberger, Fred. 1978. Summary of Evaporation in Michigan. Michigan Weather Service, East Lansing.

Penman, H. L. 1948. Natural evaporation from open water, bare soil and grass. Proc. R. Soc. Lond. Ser. A, 193:120-145.

Thompson, J. C. 1963. Weather decision making--the payoff. Bull. Am. Meteor. Soc. 44:75-78.

Vinzani, P. 1980. Personal communication.

ACKNOWLEDGEMENT

We give credit to Drs. C. R. Camp and G. D. Christenbury, Florence, SC, for furnishing the data for comparing the three methods of irrigation scheduling and to Milton Brown and John Purvis, South Carolina Weather Service, Columbia, SC, for furnishing the 5-day forecast in this study.

IRRIGATION SCHEDULING

USING WASH TUB EVAPORATION PANS

Gerald L. Westesen
Member ASAE

Thomas L. Hanson
Member ASAE

Optimum crop production under irrigation requires application of the proper amount of water at the proper time. This is a simple and teachable concept. Field applications of irrigation water, however, are often not timely or in proper amounts. Numerous irrigation scheduling devices and procedures have been used to assist the irrigator in attaining proper water application. Methods such as tensiometers, electrical resistance blocks, gravimetric soil sampling, soil feel and appearance, pan evaporation, both simple and complex formulas, and computer models have been evaluated and discussed by researchers, but have gained wide acceptance by farmers only in areas of intensive irrigation development, expensive water or energy, or high valued crops.

Evaporation data has been collected for many years from standard evaporation pans at most climatological stations. The evaporation from a free water surface integrates many climatic factors that influence evapotranspiration by crops. These factors include solar radiation, temperature, wind and humidity. Thus, measurement of evaporation can give individual applicators an indication of plant water use in the field and assist in determining when to irrigate and how much water to apply.

Research has shown that a No. 1 wash tub can be used to acceptably measure evaporation (Sims and Jackson 1971). This tub is about 48 cm in diameter and 25 cm deep with slightly sloping sides. The field installation includes a wire screen top to keep out animals, tie down stakes, and wooden blocks to keep the tub off the ground. The total cost is about $10, compared to $500 for a standard pan and accessories. The recommended installation procedure is not complicated (Westesen 1978) and is shown in Fig. 1. Other non-standard containers such as oil drums should not be used unless calibrated.

Fig. 1. Evaporation Wash Tub Field Installation

1/ Professors of Agricultural Engineering, Montana State University, Bozeman, Montana; "Contribution from Agricultural Experiment Station, Montana State University, Bozeman and Published as Journal Series No. 1216."

Comparison of pan evaporation and crop potential evapotranspiration depends upon crop grown, stage of growth, and available soil moisture. The ratio of potential evapotranspiration to pan evaporation is called the crop coefficient. The seasonal crop coefficient is approximately 1.00 for alfalfa and 0.90 for wheat (Hargreaves and Christiansen 1966). Coefficients will also vary during the season. Small grains have a coefficient of 0.33 at 25% of the season and 0.90 at 65%. A crop like alfalfa at northern latitudes establishes complete ground cover quickly in the spring with a coefficient of 0.75 achieved 25% into the growing season. With a 100-day season, this is less than one month. For practical irrigation scheduling purposes in areas of low cost water and low value crops an assumption of a coefficient of 1.00 is not unreasonable. Using a crop coefficient of 1.00 for the entire growing season eliminates one significant variable when a farmer is using a tub for scheduling purposes.

The net annual irrigation requirement for alfalfa in Montana is about 45 cm, and for small grains about 28 cm (Christensen and Westesen 1978). With alfalfa, if a crop coefficient of 1.00 was used for the entire growing season and scheduling done with pans, there would be an over-irrigation of about 8 cm for the season. This over-irrigation would occur in the spring and fall. Depending on soil moisture at irrigation, the excess 4 cm in the spring could help build up deep soil moisture reserves. In the fall, the excess would be stored if some deficit irrigation occurred over the summer. Root zone storage capacity is flexible and the root zone is not well defined, so the "over-irrigation" would not always be lost. In a state like Montana where excess irrigation often amounts to meters of water, a total seasonal over-irrigation of 4 to 8 cm would be a desirable management achievement.

The picture is not as straightforward with small grains. Over-irrigation would be about 13 cm using a crop coefficient of 1.00. Here judgment must be used in reducing irrigation somewhat below tub evaporation early and late in the season. Some over-irrigation may not be all bad. The root zone should be at field capacity at the boot stage and using an old golfing term "never up, never in", field capacity cannot be achieved by under watering. Having too much water available late in the season can cause protein dilution on wheat and possibly be costly, but the wash tubs can still provide a guide to the amount of evapotranspiration and resulting soil moisture conditions.

Growing season precipitation is very significant in Montana and portions of other states in the Northwest. The geographic distribution of precipitation is often erratic. Where irrigated lands are scattered over long river valleys this is particularly true. Commonly one field will receive a centimeter or more of moisture while another a few kilometers away receives nothing. A great advantage of wash tubs is that several can be placed in the field actually being irrigated.

Two distinct methods of tub use are available. The first method can be used for both flood and sprinkler irrigation. The tub is placed outside the irrigated area and a rain gauge is used to measure precipitation. Tub readings are made daily or every few days. A soil tube or probe is used to determine soil moisture conditions in the spring. Using a bookkeeping procedure, soil moisture depletion is subsequently determined from pan evaporation and rainfall. A problem with this method is determining what portion of rainfall is effective. The authors have recommended that rains of less than 3 mm be ignored and that 75% of any additional rainfall be considered effective (Westesen 1978). This appears reasonable, but has not been verified by research. Field trials in 1980 indicated that over several days of light rainfall with cool and humid conditions rainfall of less than 3 mm was effective in increasing soil moisture (Wallace 1980). The chore of daily pan and rain gauge readings and the requirement of bookkeeping has led to poor acceptance of this method by farmers. In addition, if the irrigation water is not measured, the values obtained can only be used for timing of irrigation, not for determining proper application amounts.

Field trials were conducted in 1980 on the Fairfield Bench near Great Falls, MT (Wallace 1980). Irrigation scheduling was done by two methods, tubs and the Irrigate Program, part of the AGNET computer network. The Irrigate Program uses climatic factors to estimate evapotranspiration which is considered equivalent to soil moisture depletion. Soil moisture depletion is also measured when possible with resistance blocks. Irrigation scheduled with the tubs resulted in a soil moisture level in the root zone as measured with resistance blocks within 2.5 cm of that estimated with the Irrigate Program. The question as to which estimation method was most correct was not answered. In the authors' opinions, a 2.5 cm variation is acceptable. Representative values for root zone depth, soil depth and available water holding capacity are difficult to quantify to the degree of accuracy implied by some irrigation scheduling techniques.

The second method of using the tubs applies only to sprinkler systems. With this method, the tubs are placed in the field and both irrigation and rain water are collected. Two reference marks are established on the tubs or a gauge is used. The reference marks are separated by an amount equal to the allowable soil moisture depletion or the net application made by a center pivot. The irrigator operates the sprinkler system so that the tub water level is maintained near the top mark. The tub functions as a rain gauge and catches all precipitation as well as irrigation water. Excess water application can still occur. For example, a heavy rain can immediately follow an irrigation that filled the root zone. If this situation occurs, the tub should be emptied to the top reference mark which accounts for excess water draining out of the root zone.

When using a center pivot a normal individual application might be about 2.5 cm. The authors recommend that three tubs be set along a radius line where they can be easily observed. The irrigator starts the system when the water level has dropped 2.5 cm from the top mark. The system passes over the tubs as it makes one revolution. When the line of tubs is approached again, the irrigator decides whether to continue irrigation or to shut down until the water level in the tubs drops to the lower reference mark.

This method has been well received by farmers. Farmers can see the water level drop in the tubs and do not have to relate water use to some esoteric formula. Many have been amazed as to how much water actually evaporates. Tubs have been successfully used for three years in the Townsend, Helena, and Chinook areas of Montana and their use is increasing. Farmers report increased yields and feel that they are irrigating correctly.

A three tower center pivot was operated by the Agricultural Engineering Department for two years near Bozeman, at an elevation of 1480 m. Only one wash tub was used under the sprinklers for irrigation scheduling because of the system size. Alfalfa (Ladak 65) was grown on the sandy loam to sandy clay soil. The area is cool and windy with nighttime temperatures often dropping to freezing in mid-July. Climatic data for 1980 is shown in Table 1. The two years of record were similar.

Table 1. Climatic Data for Red Bluff Ranch - April 25 to October 17, 1980

	Range	Mean
Temperature, Maximum	5 to 34°C	22°C
Temperature, Minimum	-5 to 20°C	6°C
Relative Humidity		34%
Wind, meters/day		1483

The tub used under the system functioned as an evaporation pan, rain gauge and catch can. Decisions on when to irrigate were based on the water level in the tub. Cumulative amounts of water applied by irrigation plus that received as precipitation during the irrigation season are shown in Fig. 2 and 3 for the two years of record. In determining the amount of water to apply, it was assumed that crop water use was zero during the harvest period. The amounts of water needed as indicated by alternative scheduling methods are also shown. The alternative methods used for comparison were a standard evaporation pan outside the irrigated area, Comp-U-Water (Valmont Industries 1978), and adjusted Blaney Criddle (SCS 1970).

Comp-U-Water is a scheduling method developed by Valmont Industries based on temperature and days into the growing season. Blaney Criddle is based on temperature, day length, and crop coefficient. The standard evaporation pan data is not adjusted by a crop coefficient. The yields were 9.8 and 8.4 tons/ha which is very good in this climate.

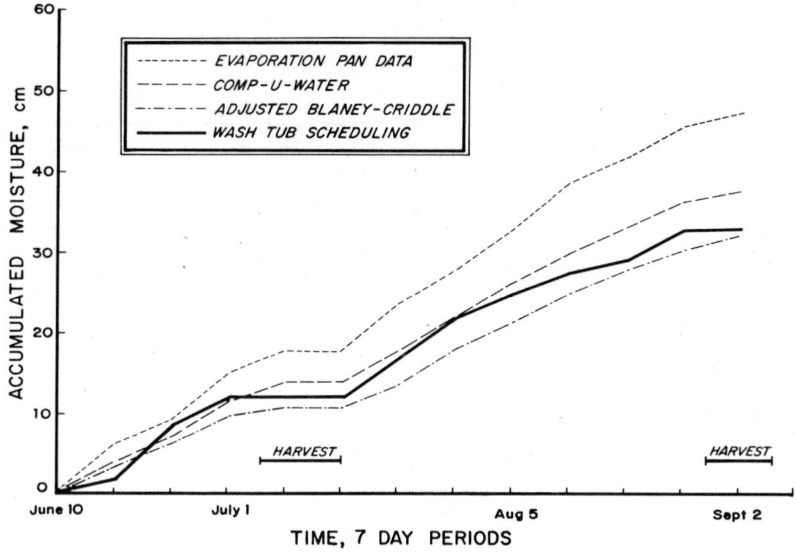

Fig. 2. Moisture Applied and Predicted Assuming No Water Use During Harvest, Red Bluff Ranch, 1979

The standard evaporation pan data shows water use markedly higher than the water use indicated by the other methods. This is at least partially attributable to the pan location outside the irrigated area.

These studies showed that Comp-U-Water, Blaney Criddle, and wash tub scheduling methods gave similar results. The difference in the tub as compared to the other two methods is about 5 cm. A partial reason for the low water application using the tub scheduling was that the system often shut down at night because of low temperatures during the peak use period. It is not possible with the information available to say which method gives an absolutely correct measure of water use, and probably none of them do.

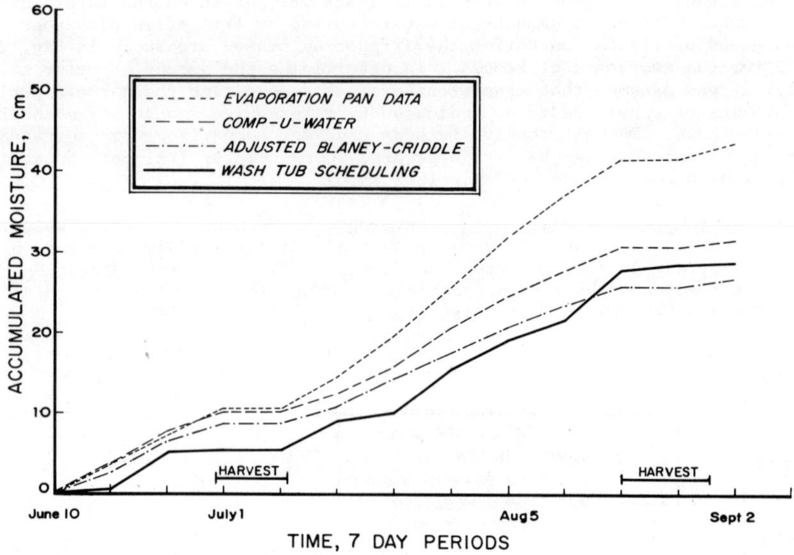

Fig. 3. Moisture Applied and Predicted Assuming No Water Use During Harvest, Red Bluff Ranch, 1980

The evaporation tub scheduling technique is simple and uncomplicated. The method can be used by farmers, extension personnel, sprinkler dealers, consultants, and manufacturer's representatives. The method is inexpensive, troublefree and requires little maintenance. The wash tub technique is particularly adapted to the varied soil types and scattered agricultural areas common in Montana and other areas of the Pacific Northwest and should be carefully considered when irrigation scheduling programs are contemplated.

REFERENCES

1. Christensen, N.W. and G.L. Westesen. 1978. Irrigation--When and How Much. Folder 172. Montana Cooperative Extension Service, December 1978.

2. Hargreaves, G.H. and J.E. Christiansen. Consumptive Use Calculations from Evaporation Pan Data. Methods for Estimating Evapotranspiration. Irrigation and Drainage Speciality Conference. American Society of Civil Engineers. November 2-4, 1966, pp. 35-64.

3. Soil Conservation Service. Irrigation Water Requirements. Tech Release No. 21. USDA-SCS. 1970.

4. Sims, J.R. and G.D. Jackson. 1971. Field Measurement of Pan Evaporation. AGRONOMY JOURNAL Vol. 63, March-April 1971, pp. 339-340.

5. Wallace, J.C. Irrigation Scheduling Demonstration on the Fairfield Bench. Unpublished Paper, Montana Cooperative Extension Service, September 1980.

6. Westesen, G.L. 1978. Scheduling Irrigation with Evaporation Pans. Circular 1211. Montana Cooperative Extension Service. August 1978.

7. Valmont Industries. Comp-U-Water. Valmont Industries, Inc., Valley, Nebraska. 1978.

DEMAND THEORY AND APPLICATION IN IRRIGATION DISTRICT OPERATION

C. M. Burt 1/
Member ASAE

J. M. Lord, Jr. 2/
Assoc. Member ASAE

For modern irrigation scheduling to be effective farmers must be able to regulate the frequency and duration of irrigations. A recent study of irrigation district operation performed by J.M. Lord, Inc. on behalf of the California Department of Water Resources, Water Conservation Division, investigated the present operating procedures used by irrigation districts covering approximately 4 million acres of land in California. Personal interviews were conducted with district managers and engineers to evaluate district water delivery flexibility, attitudes, and engineering potentials and problems. Over 100 factors such as pricing policies, field size, delivery system type (e.g., sloping canal, semi-open pipeline, closed pipeline), types of water measurement devices, numbers of employees, etc. were determined and correlated with the degree of flexibility in frequency, flow rate, and duration of water delivery offered by the district. This report will concentrate on some of the most important and least understood of these factors which affect flexibility of water delivery to the farm turnout.

Fifty seven of the 63 districts which participated in the survey claimed to deliver water "on demand". This means that farmers could receive water when they asked for it in advance. The significant point brought out was that the mean advance notice required by the districts was 32.9 hours. Another fact was that only 37 of the districts allowed the duration of an irrigation to be changed once an irrigation had begun. Among these 37 districts there again was the problem of the required advance notice, which in this case averaged 16.2 hours before a duration change could occur.

Because of operational problems which district personnel felt would occur if farmers operated the turnout gates themselves, district personnel operated gates 54% of the time. They were able to open or close these gates within an average of 1.4 hours of the prescribed time. Because of fluctuations in canal water surface levels and pipe pressures, flow rates during an irrigation event were estimated to vary an average of 4.7% from the set flow rate.

Obviously, the precision of water delivery flexibility which is necessary to implement many of the fine-tuned irrigation scheduling practices simply does not exist everywhere in the field. It is expected that this survey will provide a better understanding of what may be necessary to obtain significant improvements in both the short and long terms.

1/ Associate Professor, Department of Agricultural Engineering, California Polytechnic State University, San Luis Obispo, California.

2/ President, J.M. Lord, Inc., Fresno, California.

A significant factor is that irrigation districts are seriously limited by their suppliers. For example, the California Aqueduct and USBR projects require advance notice for water deliveries to the districts. The mean advance notice required by the suppliers was 26.3 hours for water deliveries. It is hardly surprising that the districts must in turn impose similar advance notice requirements on their users.

The question could be raised as to why even supposedly "demand" systems, such as many irrigation districts and the California Aqueduct, are operated so as to require such long advance notice of flow rate changes. Several factors seem to be involved. First, there is a lack of understanding by designers and operators of the actual requirements of modern on-farm irrigation techniques. For example, on a scale of 0-9 (9 being the highest) irrigation district managers gave a mean answer of 3.0 when asked to rate their need to improve flexibility.

The solutions to improving delivery flexibility are not simple. Each district poses some unique problems. Some districts were built many years ago and must deal with very old canals and structures. Irrigation districts are ultimately controlled and paid for by the farmers they serve, and many of the farmers have not requested improved performance. Even if an engineer or manager recognizes the need for improved flexibility, there is very little information readily available which gives concrete solutions. As part of the DWR study we did an exhaustive international literature search and compiled a bibliography which should alleviate part of this problem. Nevertheless, there are still gaps in the literature, some of which are alluded to later in this paper. A problem with much of the existing research is that there is very little literature cross referencing, indicating that those working on new water delivery control concepts are largely unaware of similar work done in the U.S. or other parts of the world.

The civil engineers who design canal and pipeline systems learn hydraulics in school, but are not exposed to either water delivery concepts (e.g., delivery schedules, and flexibility in frequency, rate, and duration) or the engineering solutions to attain flexibility. Personnel responsible for funding research have not yet recognized the need to support significant research in new water delivery methods or control logic. These deficiencies manifest themselves on the district level, where the managers were rated on their familiarity with existing devices or concepts (on the same 0-9 scale, 9 meaning extremely familiar with both the hardware and theory), and the average overall score was a low 3.1 regarding concepts presented later in this paper.

The intent of this presentation is two fold: 1) to present the argument that it is unrealistic to expect large gains in water use efficiency simply through improving or refining the existing irrigation scheduling methods without also improving the flexibility of our water delivery systems, and 2) to present a short summary of some of the possible solutions which exist today. The brief summary and comments, together with a selected list of references, will hopefully assist those who are not familiar with the concepts of water delivery control. It should be emphatically stated that research support will be needed to develop new concepts and modify, implement, and evaluate existing but non-applied knowledge and techniques.

Upstream Control - Standard Operation

Most canal systems are operated using upstream control (Buyalski, 1977). Upstream control can be classified as "supply" operation rather than "demand" operation. Conventional upstream control means releasing water from an upstream source in anticipation of demand downstream. Once the water is released on sloping canal systems, the water must be used or spilled at the lower end.

With these systems, the job of the ditchriders is primarily one of trying to shuttle water around and maintain a relatively constant water level at the turnouts. If a user shuts off his water there is no way to shut off the water coming from upstream short of closing each check gate in succession, starting with the gate at the water source. For a ditchrider several miles from the source this is difficult if not impossible. Because of transient flow hydraulics it is also difficult for a ditchrider to adjust gates properly if there are many flow rate changes. The result is that users are forced to take water deliveries for a preset number of hours simply to keep the canal water levels stabilized.

Traditional upstream control requires farmers to give water orders several days in advance. Water designated for a particular farm may need to be released from the source 3 days in advance because of travel time. Gates are operated one at a time to handle waves or large changes of flow coming down the canal system. In order to adequately meet all demands, a little extra flow can be put in at the source and spilled at the far downstream end of the canal system. But in systems where spilling is critical a district will include shortages in the operational mode. These are often referred to as "carry overs" which must be allocated to the water users. In general this results in delaying an irrigation delivery by one day.

Automated Upstream Control-General (Bucheim, et. al., 1980, Buyalski, 1970, 1977 ,1979(2), Corriga and Patta, 1979, Corriga, et. al., 1980, WSAT, 1973)

An "automated system" can imply many things and typically means some type of automatic gate control to maintain a constant water level upstream of the gates. Automatic upstream control makes the job of the ditchrider considerably easier and water levels at the turnouts can be controlled much better than with manual operation. The basic principle of upstream operation remains unchanged, however. If the upstream turnout demand increases, the downstream gates automatically shut down to maintain the constant water level. The result is that the users on the far downstream end of the system do not receive enough water. A decrease in upstream demand has the opposite effect. The gates automatically open to allow the extra flow to pass by. This flow is then ultimately wasted at the lower end of the canal system.

It may be noted that for normal operation of a canal system, rapid flow fluctuations may only be 10-20% of the average. Therefore, it may not always be necessary to automate all the gate structures. Check structures are often comprised of parallel gates, only one of which may need to be automated to provide the control need for normal flow changes. For larger flow changes, the other gates can be manually operated.

There is some confusion between remote control and automation. Gate and canal control from a remote site is not the same as an automated system. Some remote control devices are limited in their adaptability to a flexible system management program.

Neyrpic AMIL Gates - Upstream Automation on Sloping Canals (Alsthom Atlantic, 1980)

AMIL radial gates have an integral float positioned on the upstream water surface, maintaining a constant water level there. These gates are very simple mechanically and require no external power supply. They do require periodic maintenance such as minor lubrication. Concrete formwork required around the gates must be made to rather exact specifications. Installation should be done with factory supervision.

Little Man Controller - Upstream Automation With Float Controlled Electric Motors on Gates (Buyalski, 1970, WSAT, 1973, Williams, et. al., 1980)

The Little Man Controller is a simple and reliable device which was developed by the USBR. A good basic description of the various control modes required in even such simple devices as the Little Man is found in the report "Water Systems Automation," a state of the art report by the USBR in 1973.

The sensor mechanism of the device consists of a float, tape, pulley and counter-weight assembly. Two micro-switches, for upper and lower limits, are tripped mechanically either by cams attached to the float pulley shaft or by a tripping arm attached to the float tape. The micro-switchs activate timers which only energize for a few seconds at a time to raise and lower relays to the gate motor controller. Additional timers and microswitches can be installed to have different gate speeds depending upon the deviation of the water surface from desired. An "antihunt device" can be installed to deactivate the gate movement when the water level stops moving away from the set point and begins to return towards it.

Dynamic Regulation (Clemmens, 1979, Cunge, 1966, S.C.P.)

Dynamic regulation generally uses some method of statistical estimation of anticipated demand for water. Water is released into the canal network to match this anticipated demand and flow rates are remotely monitored from a central location to determine if the estimates were correct. A hydraulic model of the canal system is combined with these measurements and a schedule of anticipated demands in a computer program is used to calculate the next gate movements.

The dynamic regulation concept has been successfully used, largely as a result of tremendous research and engineering effort. A major drawback to this method is the necessity of knowing all the hydraulic parameters of the canal system. Parameters such as gate discharge coefficients are difficult to assess and friction factors can change during an irrigation season.

Upstream Control on the California Aqueduct (Dewey and Madsen, 1976, Falvey and Luning, 1979, Fredericksen, 1969)

The California Aqueduct is essentially a series of level-topped pools. Flow changes are initiated by knowing in advance what the users want in the way of water deliveries. A computer calculates the anticipated required flow rates in each section of canal. A second program calculates gate openings, simulates the canal operation with these spacings, and then readjusts the calculated gate openings. These calculated gate openings are used during the day and readjusted periodically.

The California Aqueduct uses timed gate operation. Originally all check gates were operated simultaneously, but this created large hydraulic transient flow problems. With timed gate operation, a variety of gate movements can be made to suit the operating situation. One or all of the gates at each check structure can be operated and timed to start and stop several times before reaching the predetermined gate position. Because water flows from all upstream pools into all downstream pools almost simultaneously, the distance from the ultimate water source to the Aqueduct diversion point has no appreciable effect on the time needed to meet the delivery change.

Downstream Control - General Description (Mahmood and Yevjevick, 1974, Replogle and Merriam, 1980, Williams, et. al., 1980)

Downstream control refers to regulation of canal gates based on downstream water level. Therefore, downstream control often is used synonymously with the term "demand delivery." This type of delivery is the most desirable agronomically. It also simplifies delivery schedules. A downstream controlled system will automatically respond to the opening and closing of farm turnouts. A primary advantage is that farmers can shut their turnouts off without giving advance notice.

Downstream Control - Level Top Canals (Morant, 1975)

The Neyrpic Company has designed and supervised installation on numerous large projects with level top canals. A float rests on the downstream water surface and controls the setting of an AVIO radial gate. Wave speed dictates the maximum design length of each pool. Although the canal top is level, the canal floor may be sloped or level. If there is a discharge out the downstream end of a pool, the water surface will lower at that end, causing the necessary hydraulic gradient for water flow in the canal. When the wave reaches the upstream end of a pool, it causes the float to move, thereby changing the radial gate position.

An attractive aspect of downstream control is that flow rates in the canal system do not need to be known. This simplifies the operation considerably. A disadvantage of level top canals is that the downstream ends must have higher banks than might be necessary for sloping canals.

Another disadvantage is that most canal systems have turnouts for deliveries just upstream of the check gates. With downstream control found in level top canals, water fluctuations are largest at the downstream end of each pool. This problem can be partially solved by using turnout devices which have a discharge coefficient that varies depending on the upstream head, delivering a relatively constant flow rate regardless of the upstream head.

Downstream Control - EL-FLO plus reset (Buyalski, 1979(1), 1979(3), 1979(4), Buyalski and Serfozo, 1979, Schuster and Serfozo, 1972, Shand, 1968, 1971)

The USBR has developed and tested an electronic device which senses water levels at the downstream end of a sloping canal section and then controls the appropriate gate action for the gate on the upstream end. The EL-FLO device has three parameters which provide primary control action during unsteady-state flow conditions that occur immediately after a flow change downstream. These three are:

a. The electrical filter time constant
b. The water level offset
c. The proportionality factor, or gain, of the gate movement.

The complexity and difficulty of using EL-FLO is rooted in determining these factors. Essentially, it is desirable to have very fast response to changes in water level. However, a quick response with EL-FLO may cause it to over-correct and flow instabilities will result. To determine the time constant and gain, a complicated computer program must be run to closely analyze the hydraulics of a complete canal system. This is very difficult, if not impossible, because of unknown gate constants, changing friction factors, wind wave effects, etc. Although the EL-FLO method is very complicated, its development has resulted in a significant increase in knowledge about open channel transient flow control.

Downstream Control Improvement Research

As stated before, downstream control is very desirable agronomically and because canal network management is simplified. However, the level top canals are impractical in many areas, especially regarding conversion of existing systems. The EL-FLO method has been under development for almost 10 years and has yet to be successfully implemented for an entire season.

Research by the senior author is in progress to develop the logic for a new method which will adapt to sloping canal systems with a fast response and very little hydraulic modeling will be needed for implementation. The logic utilizes multiple water level measurements along each canal reach which are feed into a microcomputer which then dictates upstream gate movements. The multiple measurements and continuous scanning action give a constant update on conditions along each reach. This eliminates the need for extensive hydraulic modeling to predict consequences of gate movements. In contrast, EL-FLO depends upon a single measurement and extensive modeling must be done to estimate what is happening at other points between the canal gates. Furthermore, quick response means fairly minor changes in water surface elevations will occur. This assures constant discharges at delivery turnouts.

There are two factors in the new logic which make it highly desirable. First, flow rates do not need to be measured in the main canals. Rather, the system can respond to changes in water surface elevations, which are easier and less expensive to monitor. Second, each pool can be operated by a separate but identical control program. A central program could be used to uniformly lower water surfaces in anticipation of water releases at a point for power generation.

Combination Control

It is desirable to have the capability to deliver water to farm turnouts on a demand schedule, but it is not necessary to have downstream control structures throughout a complete canal system to achieve this. It may be much more economical to utilize upsteam control structures on the upper 2/3 of a system, install a regulating reservoir at this point, and use downstream control in the last 1/3 of the system.

Whether a system is operated on an upstream or downstream control logic, the flow rates in the upper portions of the project will be approximately equal. Flow rates at the project source can be adjusted once or twice a day to balance the water level in the downstream regulating reservior, thereby accommodating existing upstream control structures in the upper portions of the project.

Regulating Reservoirs

A key element in many simple but flexible delivery systems is the regulating reservoir. In the previous paragraphs it was noted that proper use of a regulating reservoir will allow a demand schedule to be implemented in a largely upstream-controlled system.

The two important elements to regulating reservoir design are location and size. It must first be emphasized that a purpose of a regulating reservoir is to allow easier control upstream. A regulating reservoir does not have to act as a storage reservoir, as commonly perceived. Because regulating reservoirs can buffer imprecise upstream adjustments, the obvious proper location for them is in the downstream reaches of a canal system. This is contrary to popular notions that regulating reservoirs must either be at the extreme upstream or downstream ends of a canal network.

The second element of regulating reservior design is the necessary capacity. Again, popular belief holds that these reserviors must be very large in capacity. Understanding the purpose of the reservoirs indicates that they only need to hold the fluctuation volume of one or two days of system operation. In flat topography where no pumps are used and water must flow into and out of the reservoir by gravity, this translates into a very large surface area because of the availability of only one or two feet of fluctuation elevation. An alternative may be an elevated regulating reservoir that pumps can fill and which can be emptied by gravity flow (to better accommodate downstream control below the reservoir). The added expense of pumping can be more than offset by the decreased storage area and the improved efficiencies of irrigation due to more flexibility delivery schedules.

Pipeline Systems

During the irrigation district survey the comment that "We have a pipelined system, so you can't get any better" was frequently heard. Nevertheless, it was observed that some of the greatest needs were in pipeline districts. Generally the solutions appear to be in the form of better pressure, surge, and flow rate controls. This paper will not cover pipeline system control methods because of the lack of space.

Canal and Pipeline Sizing for Demand Delivery Systems

The literature contains very little information regarding field verification of required distribution network sizes for irrigation systems operating on a demand basis. This may be due to the very few demand systems operating. The classical theoretical work in this area was done by R. Clement (Clement, 1955, 1965)

REFERENCES

1. Alsthom Atlantic, Inc. 1980. Catalog and technical description of automatic canal gates. Grenoble, France.

2. Buchheim, J.F., L.F. Ploss and L.A. Brower. 1980. Water Systems Management Users Manual. Report No. REC-ERC-80-3. USBR. Denver, Colo. pp. 1-6.

3. Buyalski, C.P. 1970. Basic Equipment in Automatic Delivery Systems. A paper presented at the November 10-13, National Irrigation Symposium, held in Lincoln, Nebraska.

4. ----------. 1977. Study of an Automatic Upstream Control System for Canals. Report No. GR-78-4. USBR. Denver, Colo. 70 p.

5. ---------. 1979. Automatic Downstream Control of the Corning Canal Check Gates by the EL-FLO Plus Reset Control System. A paper presented at the May U.S./U.S.S.R. Symposium on Automation and Remote Control of Water Resource Systems, held in Moscow, U.S.S.R.

6. ---------. 1979. Automatic Upstream Control System for Canals. A paper presented at the September 26-29, Eighth Tech. Conf., U.S. Commission on Irrigation, Drainage and Flood Control, held in Phoenix, AZ.

7. ---------. 1979. Automatic Downstream Control for Canal Headworks Pumping Plants. A paper presented at the Sept. 26-29, Eighth Tech. Conf., U.S. Commission on Irrigation, Drainage and Flood Control, held in Phoenix, AZ.

8. ---------. 1979. Mathematical Model Simulation of Canal Systems and Control Algorithms. A paper presented at the May U.S./U.S.S.R. symposium "Automation and Remote Control of Water Resource Systems," held in Moscow, U.S.S.R.

9. --------- and E.A. Serfozo. 1979. Electronic Filter Level Offset (EL-FLO) Plus Reset Equipment for Automatic Downstream Control of Canals. Report No. REC-ERC-79-3. USBR. Denver, Colo. 145 p.

10. Clemmens, A.J. 1979. Control of Modified Demand Irrigation Distribution Systems. Presented at the July 17-20, ASCE Irrigation and Drainage Division Speciality Conference, "Irrigation & Drainage in the Nineteen Eighties," held in Albuquerque, New Mexico.

11. Clement, R. 1955. Le calcul des debits dans les canalisations d'irrigation. Journees de'etudes sur l'irrigation publiees par l'Association Amicale des Ingenieurs du Genie Rural. (French)

12. --------. 1965. Calcul des debits dans les reseaux d'irrigation fonctionnant 'a la demande'. La Societe du Canal de Provence et d'Amenagement de la Region Provencala, Le Tholonet, 13. Dec. 92 p. (French)

13. Corriga, G. and F. Patta, et.al. 1979. A Mathematical Model for Open-Channel Networks. Applied Math. Modeling 3:51-54.

14. --------, S. Sanns and G. Usai. 1980. Frequency Response and Dynamic Behaviour of Canal Networks with Self-Leveling Gates. Applied Math. Modeling 4:125-129.

15. Cunge, J.A. 1966. Comparison of Scale-Model and Mathematical Model Test Results on Translatory Waves in the Oraison-Manosque Power Canal. Translated from the French Journal: La Hoville Blanche, No. 1. Theodor Strelkoff ed. An in house translation for the California DWR.

16. Dewey, H. and W.R. Madsen. 1976. Flow Control and Transients in the California Aqueduct. Journal of the Irrigation and Drainage Division, ASCE, Vol. 102, IR3:335-348.

17. Falvey, H.T. and P.C. Luning. 1979. Gate Stroking. Report No. REC-ERC-79-7. USBR. Denver, Colorado. pp. 1-15.

18. Fredericksen, H. 1969. Design of the California Aqueduct. Journal of the Irrigation and Drainage Division, ASCE, Vol. 95, No. IR2:364-386.

19. Mahmood, K. and V. Yevjevick. 1974. Unsteady Flow in Open Channels, Vol I and II. Proceedings of the Institute on Unsteady Flow in Open Channels held on June 17-28, at Colorado State University. Water Resources Publications, Fort, Collins, Colorado. Chapters 4, 7 and 13.

20. Morant, R. 1975. L'amenagement hydraulique et agricole du bassin de la Durance. Le Canal de Provence," Eau amenagement Reg. Provencale 6:1-21. (French)

21. Replogle, J.A. and J.L. Merriam. 1980. Scheduling and Management of Irrigation Water Delivery Systems. A paper presented at the October 20-23, Second National Irrigation Symposium held in Lincoln, Nebraska.

22. Schuster, J.C. and E.A. Serfozo. 1972. Study of Hydraulic Filter Level Offset (HyFLO) Equipment for Automatic Downstream Control of Canals. Report No. EC-ERC-72-3. USBR. Denver, Colo. 15 p.

23. Shand, M.J. 1968. Final Report-The Hydraulic Filter Offset Method for the Feedback Control of Canal Checks. Technical Report HEL-8-3. University of California, Berkeley.

24. --------. 1971. Automatic Downstream Control Systems for Irrigation Canals. Technical Report HEL-8-4. Univ. of Calif., Berkeley. 159 p.

25. Societe du Canal de Provence. (no date) The Canal de Provence, an Example of a Multi-Purpose Water Supply Scheme in the South of France. An informative brochure by the Societe du Canal de Provence, France. 10 p.

26. Societe du Canal de Provence. (no date). Dynamic Regulation, An Original Management Process for Old and New Canals. An informative brochure by the Societe du Canal de Provence, France. 13 p.

27. Water Systems Automation Team. 1973. Water Systems Automation-State of the Art. An unnumbered pub. of the USBR, Denver, Colo. 58 p.

28. Williams, H.R., et al. 1980. Water Systems Automation and Management, Session 2. Session notes of the 1980 Water Systems Management Workshop of the USBR held in Denver, Colorado.

HOW SCHEDULING FITS IN THE IRRIGATION PROGRAM IN THE GRAND VALLEY OF COLORADO

Brice E. Boesch Allan S. Humpherys David Young

The Grand Valley is an irrigated area of about 26,000 hectares (65,000 acres) adjacent to the Colorado River near Grand Junction, Colorado (Soil Conservation Service 1977). Grand Junction is near the Colorado-Utah border due west of Denver, Colorado.

Implementation of the on-farm Grand Valley Salinity Control Program was initiated in January 1979. The program is authorized under Title II of Public Law 93-320. The purpose of this law is "to authorize certain works in the Colorado River Basin to control the salinity of water delivered to users in the United States and Mexico." Title I of Public Law 93-320 authorized certain works of improvement below Imperial Dam in southwestern Arizona. Some of these works of improvement have been completed or are under construction (Strand et al 1981). Title II of Public Law 93-320 authorized construction of four upstream projects on the Colorado River and studies on twelve additional projects. The Grand Valley Program is one of the four authorized projects. About 1.7 million dollars per year of U.S. Government funds have been used since 1979 to cost-share the installation of on-farm irrigation systems.

The Colorado River is one of the most used rivers in the world. The water supply is utilized in seven Colorado River Basin states for public water supplies, irrigation, and energy development.

The downstream states of Arizona and California have used most of their water entitlement, while the Upper Basin states are still developing their water rights. Each new irrigation project has the potential of increasing the salt content of the Colorado River because of the salt in the substratum below the irrigated land. Subsurface return flows dissolve this salt and carry it back to the Colorado River. Also, increasing irrigated acreage increases consumptive use and thereby increases salinity concentrations. Transmountain diversions in the upper reaches of the Colorado River also affect its salinity level. The water diverted in the mountains is quite pure; after diversion it is not available to dilute downstream Colorado River water supplies.

The purpose of Title II of Public Law 93-320 is to keep the salt loading of the Colorado River from increasing as water development is increased. This objective is being met through work in existing projects such as the Grand Valley Program.

The Grand Valley contributes 700,000 Mg (630,000 tons) of salt per year to the Colorado River. Of this total salt load, on-farm ditch seepage contributes

The authors are: Brice E. Boesch, Water Management Engineer, USDA, SCS, Denver, Colorado; Allan S. Humpherys, Agricultural Engineer, Snake River Conservation Research Center, ARS, USDA, Kimberly, Idaho; David Young, Agricultural Engineer, Colorado State University, Grand Junction, Colorado

145,000 Mg (130,000 tons) and on-farm deep percolation contributes 125,000 Mg (112,000 tons)(Keys 1981). Each dam^3 (0.8 acre-foot) of seepage from ditches or deep percolation of water below crop rooting depths carries about 3.7 Mg (3.3 tons) of salt back to the river in underground irrigation return flows.

AGENCY COOPERATION

The Agricultural Stabilization and Conservation Service (ASCS) provides cost-sharing to farmers through the Agricultural Conservation Program. The Soil Conservation Service (SCS) provides technical services. The Agricultural Research Service (ARS) provides technical expertise in several areas including monitoring of the installed on-farm systems and the semiautomation of surface irrigation systems. The Bureau of Reclamation (USBOR) assists in the monitoring activities and improves the canals and some of the laterals by installing concrete lining or pipelines. The Mesa Soil Conservation District (SCD) has taken the lead in information activities and in helping to shape local programs. Colorado State University (CSU) is assisting in the monitoring program and in ongoing research (Strand et al 1981).

ON-FARM PROGRAM

Water Management

Water seepage from on-farm ditches can be reduced by lining the ditches with concrete, installing gated pipe systems, and/or underground plastic pipelines to carry the required irrigation water.

Deep percolation of the irrigation water below the crop root zone can be reduced through effective water management techniques. Good Water Management has two basic requirements:

A. An irrigation system that can deliver irrigation water uniformly to the crop in the right quantity and at the right time.

B. An irrigator who knows and follows good water management principles and practices including:

   1. How to determine when water should be applied, based on the rate of water use of crops and on the stages of plant growth.

   2. How to measure or estimate the amount of water required for each irrigation, including the leaching needs.

   3. The normal time needed for the soil to absorb the required amount of water and how to detect changes in the intake rate.

   4. How to adjust stream size, application rate, or irrigation time to compensate for changes in such factors as intake rate or amount of water to be applied.

   5. How to recognize erosion caused by irrigation.

   6. How to estimate the amount of irrigation runoff.

   7. How to evaluate the uniformity of water application (Soil Conservation Service 1973).

Soils in the Grand Valley are generally fine textured, such as silt loams and silty clay loams. Typically, furrow and corrugation irrigation systems are used. Sprinkler irrigation has been evaluated in Grand Valley and is presented to landowners as an alternative irrigation system. Very few sprinkler systems are actually installed, however, because of high energy costs and soil characteristics. The soils of the Grand Valley are heavy enough to cause runoff from sprinkler irrigated land unless special designs are adapted to the site.

Soils in the Grand Valley do not correspond to the standard SCS intake families. Evaluations, discussed elsewhere in this paper, are being conducted to provide intake values that will be representative of Grand Valley soils.

Most farmers in the grand Valley also have jobs off their farms. Thus, the present practice is to change irrigation sets in the early morning or in the evening because of the inconvenience of changing sets at night or during midday. The result is 12-, 24-, or 36-hour set times.

The furrow equations used by SCS typically require 16- to 18-hour sets to apply a 7.5- centimeter net irrigation. The common practice of irrigating longer than is needed causes excessive deep percolation and salt loading in the Colorado River.

Semiautomated Irrigation Systems

Semiautomated surface irrigation systems allow flexibility in the set time because the farmer does not have to be present to change the water. The semiautomated equipment is generally quite simple. Mechanical and battery-powered time clocks are used to control equipment because electricity is not available in remote field locations. The most commonly used semiautomated surface irrigation systems are the ported ditch system and the gated pipe system.

Ported Ditch System: The ported ditch system is preferred by some farmers because of its permanence and ease of operation. Set widths are determined from SCS furrow equations and the farmer's historical water supply. Generally a flow rate of 0.6 to 0.9 liters per second (10 to 15 gallons per minute) is needed for each 75-centimeter (30 inch) spaced furrow. Typical water supplies vary from 0.03 to 0.09 cubic meters per second (1 to 3 cubic feet per second). Therefore, set widths vary from 25 to 110 meters (75 to 330 feet), widths of 25 to 45 meters are the most common.

The ported ditch system consists of a concrete-lined ditch that "stair steps" across the top of the field. Each section of ditch is installed at a uniform grade and its length corresponds to the width of an irrigation set. A grade of 0.10 percent is used to prevent sedimentation. Holes are drilled through the concrete lining and ditch berm, and 5-centimeter (2 inch) diameter plastic tubes are installed about 15-centimeters (6 inches) below the top of the ditch. The plastic tubes are long enough to protrude through the ditch bank and into the field area. An adjustable cover is installed on the tube inlets to provide flow adjustments through the ports. All ports are installed at the same elevation for each irrigation set.

The installation error in port elevations caused by mechanical difficulties in drilling through the concrete ditch lining is $\pm$ 0.75-centimeter ($\pm$ 0.25 inch) from level. Furrow flows are adjusted the first time the system is operated in the spring. The rate of advance of water down the furrow is observed, and the furrow stream size is adjusted to obtain as uniform a rate of advance as possible in all furrows. Trash plugging of ports does not appear to be a large problem.

"Drop-closed" or "drop-open" gates are installed at the downstream end of each ditch section. The normally open "drop-closed" gate is the most common ditch check used. It is equipped with a timer that releases the gate at a predetermined time. The gate drops closed and moves the irrigation water one set upstream.

A minimum drop of 10 to 15 centimeters (4 to 6 inches) is installed at the downstream end of each ditch section. Thus, when the ditch check located just upstream of this drop is opened, the water surface is below the ports in an upstream section while flowing out of ports in a downstream set.

The normally closed, "drop-open" gates are used where irrigation is to start at the upstream end of the field. At the required set time the timer allows the gate to open and the irrigation head flows downstream to the next irrigation set, where the gate is closed.

Gated Pipe System: Another type of semiautomated irrigation system commonly used is the "flow-through" gated pipe system. Gated pipe is installed in sections on a "stair-step" grade similar to the ported ditch system.

A timer-controlled valve is installed in each section of pipeline corresponding to an irrigation set width. The timer is mounted on top of the check valve. The check valve, built by a local Grand Valley manufacturer, operates in the pipeline similarly to a "drop-closed" gate in a ported ditch. When released by a timer, the check valve closes and moves the irrigation water one set upstream. This type of valve has the disadvantage of causing a condition similar to water hammer in the gated pipeline. As the check drops into the water surface the water velocity slams it shut, causing the gated pipe to separate or rotate.

A spring-actuated butterfly valve has been developed by the second author and Grand Valley manufacturers to take the place of the check valve. This valve is also operated by a 24-hour mechanical timer. At a preset time the timer releases the butterfly valve and allows the irrigation head to proceed downstream to the next irrigstion set. This valve eliminates the shock associated with the closing of the check valve and also costs less.

The gated pipe is placed so that the gates are at about 30 degrees from vertical. This placement allows the water to flow beneath the open gates as open channel flow for downstream irrigation sets, so the gated pipe must be about one size larger than would otherwise be used. Thus, the gated pipe operates as both a water carrier and a water distribution system.

Water flowing out of the upturned ports sometimes causes erosion unless "erosion socks" are installed to dissipate the energy of the water. Erosion socks are made of fabric and held in place by a wire fastened around the gated pipe. The water falls into the sock and flows out onto the ground without causing serious erosion although the "erosion socks" sometimes kink and cause serious erosion next to the pipeline. An alternative erosion control method involves placing 5- to 6-foot wide sheets of plastic on the furrow side of the gated pipe. The water falls onto this plastic, which dissipates the water's energy without causing serious erosion.

Pneumatically controlled valves are also used in gated-pipe irrigation sets. Portable air tanks are used to provide the high-pressure air supply. Water pressure is used to control valve changes in a commercially available valve.

"Cablegation" is a new technique being tested in the Grand Valley. In "cablegation" a polyvinyl chloride pipe is placed on a constant grade and is buried so that the top is about even with the ground surface. Orifices spaced to correspond to furrow spacing are drilled in the top of the pipe. The pipe is sized to carry the full water supply without filling the pipe to the top. A geared down electric motor controls the rate of movement of a plug on a cable inside the pipe which forces the water out of the pipe. As the plug moves downstream, the "head" on each orifice decreases, thus decreasing furrow flow rates (Kemper et al 1980).

MONITORING PROGRAM

The SCS contracted with ARS to provide monitoring services for the on-farm program in Grand Valley for calendar year 1980. This contract has been renewed for 1981. The purpose of the monitoring effort is to determine how effective the on-farm program is in reducing salt loading of the Colorado River.

The monitoring service is provided by CSU under contract to ARS. Water budgeting is used to determine the amount of deep percolation of water in irrigated fields. Deep percolation is the amount of water applied to the field minus tailwater runoff, and crop consumptive use, plus effective rainfall. Effective rainfall is precipitation falling during the growing period of the crop that is available to meet the consumptive water requirement of the crop. The leader of the Grand Valley on-site monitoring team has a desk in the SCS office. This location promotes frequent exchanges of information.

SCHEDULING PROGRAM

A SCS water-management technician was assigned to the Grand Valley on May 1, 1981. This technician helps landowners and operators to obtain proper water management on farms that have improved irrigation systems. Thirteen farmers were selected for this year's scheduling program. The basis for selection included the possession of fields with different soils, the production of different crops, and the farmer's willingness to participate in this program.

After the farms are chosen, the semiautomated equipment installed under the salinity control program is evaluated to determine that it is all in working order and is properly used. The second step is to evaluate the furrow irrigation systems. This evaluation consists of running furrow trials on several irrigations to determine intake families and how they vary from irrigation to irrigation. Information is also gathered on advance-recession time and depth to which irrigation water penetrates the soil at several locations in the field.

Soil augers and probes are used to determine the depth of water penetration after all gravity water is drained out of the soil profile. Water penetration is checked in at least five locations in each irrigation set. The locations monitored are at the center of the field and in the four corners. A distance of 30 meters (100 feet) is paced diagonally from the corner of the field towards the center. The resulting points are the approximate locations of the four corner checks. The effects of seepage from headwater and tailwater ditches on soil profile wetting are minimized at these locations. Different soils, unusually flat lands or steep slopes in the irrigation set, are also checked for water penetration.

Early research by CSU and ARS in the Grand Valley indicated that the first two irrigations of the season contribute the most deep percolation (Evans et al 1978). These early irrigations are on soils that are not yet water compacted and are, therefore, loose and easily penetrated by the irrigation water. The furrow intake rate is highest on the first irrigation and somewhat lower on the second irrigation. Later irrigations have lower intake rates than the first two but do not vary much from one another. Knowing how the intake rate varies will help SCS personnel to better design furrow irrigation systems.

The third step in the water management program is the irrigation scheduling. Proper scheduling of irrigations in the Grand Valley keeps the soil moisture level between 50 percent and 100 percent of field capacity with minimum deep percolation. Farmers can obtain maximum crop growth while reducing deep percolation to the minimum needed for the necessary leaching fraction. Farmers are being shown how to determine when to irrigate and how much water to apply to minimize excessive deep percolation.

A combination of the "bookkeeping" method and soil moisture monitoring is used in scheduling. Daily consumptive use is deducted from the maximum amount of water that can be extracted from the soil. Several days before the bookkeeping account shows that it is time to irrigate, soil moisture in a number of locations in the field is examined for verification. The account is adjusted if there is a significant difference.

SUMMARY

The Grand Valley contributes 700,000 Mg (630,000 tons) of salt per year to the Colorado River. The Grand Valley Salinity Control Program was implemented in January 1979 to help farmers improve their irrigation systems. Good water management reduces deep percolation of water below crop root zones, thus, salt pickup is reduced in subirrigation return flows to the Colorado River.

SEMIAUTOMATED IRRIGATION SYSTEMS

Fig. 1. Ported Ditch System

Fig. 2. "Drop-closed" Gate

Fig. 3. Semiautomated Gated Pipe System

REFERENCES

1. Soil Conservation Service. 1977 On-farm Program for Salinity Control. A final report of the Grand Valley Salinity Study. 69 p.

2. Strand, R.I., H.R. Pritchett, B.E. Boesch, J. W. Keys, E.G. Kruse and R.S. Swenson. 1981. Irrigation in the Colorado River Basin, U.S.A. solving the salinity problem to save a giant. I.C.I.D. Proceedings - 11th Congress on Irrigation and Drainage.

3. Keys, J.W. III. 1981. Grand Valley Grand Valley irrigation return flow case study. Irrigation Drainage Div. Proc. Amer. Soc. of Civil Eng. Vol. 107, No. IR2, June 1981.

4. Strand, R.I., B.E. Boesch and E.G. Kruse. Salinity control - The Colorado River Experience. Water Forum 81 Management Group D, Specialty Conference, ASAE, San Francisco, CA. August 10-14, 1981.

5. Soil Conservation Service 1973. Engineering Standard and Specification No. Colo-449. 1972.

6. Evans, R.G., W.R. Walker, G.V. Skogerboe and S.W. Smith. 1978. Evaluation of Irrigation Methods for Salinity Control in Grand Valley. Environmental Protection Agency.

7. Kemper, W.D., W.H. Heinemann, D.C. Kincaid and R.V. Worstell. 1980. Could Cablegation Reduce the Cost of Your Irrigation?

A COMPARISON OF VARIOUS

IRRIGATION SCHEDULING PROCEDURES WITH CORN

Paul E. Fischbach [1/]
Member ASAE

Irrigation Scheduling is applying the right amount of water at the right time. In areas where rainfall occurs during the growing season, special irrigation management procedures need to be applied in order to make effective use of the rainfall. Pilot projects in Nebraska indicated that about 178 millimeters of water per season may be saved by scheduling. If the water is lifted 30 meters out of the ground water reservoir, 109 and 251 liters of diesel fuel per hectare can be saved annually with gated pipe and sprinkler systems respectively. Excess water applied that results in deep percolation below the root zone also carries nitrate-nitrogen with it. This is also an economic loss to the irrigator. Timing of the irrigation application also has an influence on the yield of crop. A delay of four or five days at the reproductive stage of corn on a sandy soil may reduce grain yields 1245 kilograms per hectare or more.

Objective

To compare various irrigation scheduling procedures in regard to amount of water applied with grain yields.

Procedure

A 3.8 hectare field with a Sharpsburg silty clay loam soil was selected for the experiment. The Sharpsburg soil held approximately 45 millimeters of water per 30 centimeters of soil depth. The length of irrigation run was 400 meters with a slope of 0.2 percent. The water supply was from a 3420 liters per minute irrigation well delivered to the field through a 25.4 centimeter diameter pipe. The field was surface irrigated with gated pipe. The runoff water from the field was collected into a sump and pumped back into each treatment. A propeller meter was used to measure the water to the treatments. The gravimetric procedure of determining soil moisture was used at planting time and at the end of the season. The field was planted to corn (Pioneer 3388) all three years sometime between May 1 and 10 with a 0.9 meter row spacing. The average plant population was 50,000 plants per hectare. The nitrogen was applied at the rate of 178 kilograms per hectare when the corn was 45 centimeters tall. Rainfall was measured by a standard weather bureau rain gauge located near the field. The soil moisture was at field capacity to a depth of 2 meters on all plots soon after planting.

Irrigation Treatments

The four irrigation scheduling treatments were as follows:

1. Et - the determination of crop water used was modified Blaney-Criddle procedure (Schleusener and Kruse 1963). Maximum and minimum temperatures were used in calculating crop water use by the empirical formula. These

1/ PAUL E. FISCHBACH, Professor, Agricultural Engineering, IANR, University of Nebraska, Lincoln, Nebraska 68583.

plots were irrigated when the crop water use as calculated by the formula, less rainfall, was approximately 76 millimeters. Then approximately 76 millimeters of water was applied. However, the amount varied from 62 to 83 millimeters.

2. Electrical Resistance Blocks (Fischbach 1971) - Electrical resistance blocks were placed at the 15, 45, 75, and 105 centimeter depths about 30 meters from the upper and lower end of each replication. Therefore, there were 6 block stations for this treatment. The plots were irrigated when the 45 centimeter depth block indicated 50 percent of the soil water had been used. Then 50 to 76 millimeters of water was to be applied. However, the irrigations averaged 69 to 83 millimeters over the 3 years the experiment was in progress.

3. Stage of Growth (Orlando Howe 1955) - The procedure used for this treatment was to irrigate when the corn was about 76 centimeters tall (soon after furrowing), early tasseling stage, pollination (silking), and milk stage of the kernals on the ear of corn. Rainfall during the growing season influences whether the crop and/or soil can utilize an irrigation. Therefore, the "Feel Method" of determining soil moisture in the root zone was used to adjust the irrigations a few days so that there was deficit of at least 50 millimeters of water in the root zone prior to irrigation. The depleted soil water was then replaced by irrigation.

4. Irrigate every 14 days (some irrigation district's delivery schedule) - Starting when the corn was about 76 centimeters in height, an attempt was made to apply 50 to 76 millimeters of water. The actual amount varied from 52 to 83 millimeters. The influence of rainfall was also used in the procedure. The "Feel Method" of determining soil moisture in the root zone and stage growth of the corn plant was used to determine when to stop irrigations.

The replications were 10 rows wide and 400 meters long in a randomized block design. Each treatment was replicated four times. Consequently, with the check rows, the entire experiment required 200 rows of corn. The grain yields were determined by harvesting the six center rows (400 meters long) of each replication with a corn combine. The corn was weighed and corrected to 15.5 percent moisture content.

Discussion of Results

The experiment was in progress for three years, 1972, 1974, and 1977. It was not conducted during the 1973, 1975, and 1976 seasons because of lack of funds. Rainfall varied from a low of 203 millimeters to a high of 307 millimeters during the irrigation season (June 14 to September 16).

During 1972 (Table 1) no rain exceeded 42 millimeters and the rainfall was well distributed over the irrigation season. The longest period without some precipitation was nine days. In addition, the largest rainfall occurred at the beginning of the pollination period. The greatest amount of irrigation water applied was 277 millimeters in four applications with the electrical resistance block procedure of scheduling irrigations. Grain yields varied from 9586 to 9959 kilograms per hectare; however, there was no significant difference between grain yields for all scheduling procedures. The check (no irrigation) produced 8091 kilograms per hectare of corn. This treatment was significantly lower than all irrigated treatments at the 95 percent level.

During 1974 (Table 2) only 203 millimeters of rainfall occurred between June 20 and September 15. In addition, temperatures above 38 degrees Celsius occurred early in July (weather bureau records). The non-irrigated plots did not produce grain. It appeared that the corn roots could not extend themselves fast enough into the subsoil to maintain growth where there was plenty of soil moisture. The result was no grain yield and very little dry matter production. Grain yields varied from 10021 to 10768 kilograms per hectare; however, there was no significant difference at the 95 percent level statistically between grain yields for all irrigation scheduling procedures. However, the difference of 41 millimeters of more water application on the stage of growth treatment compared to the electrical resistance block treatment was due to water application efficiency. The surface soil became so dry prior to an irrigation that excess water was applied to that treatment. This is a problem for an irrigator if he waits too long between irrigations under field conditions.

During the 1977 season (Table 3), the same experiment was repeated. Rainfall June 20 through September 16 was 307 millimeters. However, only 9 millimeters of rainfall occurred during the pollination period (July 22 through August 3). Most of the rainfall occurred from August 9 through September 12 (202 millimeters). The rainfall pattern did affect the "every 14 days" irrigation treatment as it wasn't irrigated after the rains came, although it may have been stressed slightly early in the irrigation season. Grain yields varied from 9523 to 10,021 kilograms per hectare on the irrigation scheduling treatments. There was no significant difference at 95% level between grain yields for all irrigation scheduling treatments. Grain yields on the check plot were 2801 kilograms per hectare. There was a significant difference in grain yields between the check plots and the irrigated plots. The "every 14 days" treatment received one less irrigation, only 208 millimeters of water was applied compared to 302 to 312 millimeters for the other treatments. Deep percolation occurred with nearly every irrigation during the three seasons the experiment was conducted. This resulted in excess water being applied each irrigation. It would have been better to apply less water each irrigation and leave room for rainfall that occurred during the growing season. Although it was not measured in these experiments, an estimated 76 millimeters of water was lost to deep percolation from irrigation water and rainfall during each season. In order to make effective use of rainfall that occurs during the irrigation season, it is important that less irrigation water be applied than the soil will hold within the root zone. So there will be room within the root zone to hold a reasonable amount of rainfall without deep percolation occurring.

Another study is being conducted to determine irrigation management procedures for surface irrigation. In these experiments on 0.2 percent slope and irrigation length of runs of 330 to 400 meters, maximum allowable furrow streams are being used with various furrow slicking treatments. The objective is to find a treatment which would apply only 50 millimeters of water or less and distribute it uniformly over the full length of irrigation run. These experiments are in progress.

The irrigation scheduling procedures studied would assist irrigators in saving water and energy compared to data collected from the pilot projects which show 185 millimeters of excess water applied each season. However, improved irrigation management procedures could save more water, energy and nitrate-nitrogen leaching.

SUMMARY

Four irrigation scheduling procedures for irrigating corn were tested. There was no significant difference in grain yields from the four irrigation scheduling procedures tested. However, grain yields varied from 9523 to 10768 kilograms per hectare from the irrigation treatments over the three years the experiment was conducted. The corn grain yield on the check (no irrigation) varied from 0-8092 kilograms per hectare. Rainfall and other weather factors greatly influenced the check plots.

However, during one out of the three years, the "every 14 days" treatment required one less irrigation. This was due to the dates rainfall occurred. The last irrigation of the season was omitted one season because of rainfall.

All four irrigation scheduling procedures appear to be satisfactory for use by irrigators to maintain yields and save water.

REFERENCES

1. Blaney, H.F. and W.D. Criddle. 1950. Determining water requirements in irrigated areas from climatological and irrigation data. USDA, SCSTP-96.

2. Fischbach, P.E. and J.C. Steele. 1955. Field moisture test for determining when to irrigate. E.C. 55-700, Extension Service, University of Nebraska, Lincoln, Nebraska 68583.

3. Fischbach, P.E. 1965. Scheduling irrigations by electrical resistance blocks. E.C. 65-752, Extension Service, University of Nebraska, Lincoln, Nebraska 68583.

4. Howe, O.W. 1955. "When Applying Irrigation Water Timing is Important". Nebraska Experiment Station Quarterly, College of Agriculture, University of Nebraska, Lincoln, Nebraska 68583.

5. Howe, O.W. and H.F. Rhoades. 1955. Irrigation practices for corn production in relation to stage of plant development. Soil Sci. Soc. Proceedings.

6. Schleusener, P.E. and E.G. Kruse. 1963. "Empirical Formulae for Computing Water Needs of a Row Crop", TRANS. of the ASAE 6(2): 140, 141, and 144.

Table 1. Comparing various irrigation scheduling procedures on corn (1972)

Treatment	Irrigations	Net Amount of Water Applied	Grain Yield
	No.	(mm)	(kg/ha)
Et	4	264	9648[a]
Blocks	4	277	9959[a]
S of G.	3	272	9586[a]
Every 14 days	4	264	9586[a]
Check	0	0	8092[b]

Rainfall June 14-September 20 - 240 mm

[a] NOT significant at 95% level
[b] Significant at 95% level

Table 2. Comparing various irrigation scheduling procedures on corn (1974)

Treatment	Irrigations	Net Amount of Water Applied	Grain Yield
	No.	(mm)	(kg/ha)
Et	4	333	10021[a]
Blocks	4	333	10768[a]
S of G.	3	373	10208[a]
Every 14 days	4	358	10208[a]
Check	0	0	0 [b]

Rainfall June 20-September 15 - 203 mm

[a] NOT significant at 95% level
[b] Significant at 95% level

Table 3. Comparing various irrigation scheduling procedures on corn (1977)

Treatment	Irrigations	Net Amount of Water Applied	Grain Yield
	No.	(mm)	(kg/ha)
Et	5	310	9586[a]
Block	5	302	10021[a]
S of G.	5	312	9959[a]
Every 14 days	4	208	9523[a]
Check	0	0	2801[b]

Rainfall June 20-September 12 - 307 mm

[a] NOT significant at 95% level
[b] Significant at 95% level

STATE OF THE ART IRRIGATION SCHEDULING COMPUTER PROGRAM[1/]

G.J.Harrington[2/] D.F.Heermann[3/]

An irrigation scheduling program for large digital computers was developed by Jensen (1969) and Jensen et al. (1970) for the U.S. Department of Agriculture (U.S.D.A.). This program or an adaption of it is currently being used on about 400,000 hectares in the United States. The USDA Irrigation Scheduling Program uses meteorological data to calculate water use and maintains a water budget. It also forecasts the timing and amount of irrigation water necessary for optimum crop production. The modern farm manager has consulting services available to provide estimates of present soil-water status and to predict irrigation dates and amounts of water to apply to each field. Irrigation consulting firms offer computer irrigation scheduling services, however, in many cases the service is based on field measurements of soil moisture rather than using the full potential of computer procedures. Computer procedures may provide a clearer estimate of when to irrigate and how much irrigation water to apply at a lower total cost than conventional field procedures. The consultant's use of computer scheduling procedures may be more of a selling point than a mainstay of their scheduling service because the use of computers appeals to their clients.

There are a number of reasons why computer scheduling procedures are not universally used as the primary irrigation scheduling procedure by consultants. The most important reason is that accurate feedback of dates and amounts of rainfall and irrigation are difficult to obtain. Another reason is that programming and operation errors produce incorrect irrigation schedules which can discredit the consulting service in the eyes of the clients. Other problems could be that the necessary data entry into the computer system is tedious and that there is a delay in producing the schedule compared with an on-the-spot personalized field estimate.

It has been shown (Heermann, Haise and Mickelson, 1976) that computer scheduling procedures can be used to schedule field crop irrigations for an entire season without adjustment to the computer estimated soil moisture depletions. It appears, therefore, that the problem with computer estimates lies more with acquisition of correct field information and accurate entry of this information into the computer scheduling procedure than in the error associated with the estimation procedures used in the computer programs. Field verification of computer estimates are nonetheless a necessary check on the computer schedules.

An important step in the development of computer irrigation scheduling procedures is to improve the reliability of field data acquisition and data entry. A comprehensive data checking system within the computer program can assist in preventing entry errors. Data entry and scheduling procedures must be flexible but easily used by semi-skilled personnel.

This paper discusses modifications to the USDA Irrigation Scheduling Program. The program checks ranges of input data for reliability and accepts either English or metric units. The interactive program is more flexible in providing output to meet the needs of various irrigators and irrigation systems.

[1/] Contribution from USDA-ARS, Agricultural Research, in cooperation with Colorado State University.

[2/] Senior Engineer, Ministry of Agricultural and Fisheries, Lincoln College, New Zealand.

[3/] Agricultural Engineer, USDA-ARS, Fort Collins, CO.

Kincaid and Heerman (1974) adapted the scheduling program to a programmable calculator. The current version of the program, SP1, is written in BASIC II language in an interactive mode for a WANG 2200 MVP computer. (1) The structure of the program has been revised to have more than 30 subroutines which should make the program easily configured for other computer systems. The calculations are performed in metric units.

USDA IRRIGATION SCHEDULING PROGRAM

The USDA Irrigation Scheduling Program contains necessary algorithms to maintain a water budget. The water balance equation, applied to each field for each day, is

$$D_{pi} = D_{pi-1} + K_c\, E_{tr} + E_{tw} - R_i \qquad (1)$$

where E_{tr} - reference evaporation
K_c - crop coefficient (a function of crop development)
E_{tw} - additional evaporation following an irrigation or rain.
D_{pi} - depletion on day i
R_i - sum of rainfall and irrigation on day i

If R_i is larger than the previous depletion plus evapotranspiration then the depletion is set at zero.

Crop water use

The reference evapotranspiration, E_{tr} is the rate of water use by a thoroughly watered reference crop (alfalfa) with 300 to 450mm of growth. A combination equation, similar to that used by Penman (1963) has been calibrated at several locations in the United States to estimate E_{tr}.

Daily reference evapotranspiration is calculated as a function of; maximum and minimum temperature; dew point temperature; solar radiation and wind run.

The reference E_t is multiplied by a crop coefficient and a stress coefficient to obtain the actual crop E_t. The crop coefficient is currently calculated on a normalized time scale, with time from planting until full cover on a percentage basis and time after full cover as elapsed days. (For a full discussion of crop coefficients refer to J. L. Wright's paper in these proceedings.) Maximum E_t occurs when the crop reaches effective cover. It would occur when the leaf area index is approximately equal to 3. Corn for example would reach effective cover approximately 10 days after tasseling or at a leaf area index of 3. The stress factor is:

$$K_s = \frac{\log(\ 1 + 100(\ 1 - D_P/D_T)\ )}{\log(101)} \qquad (2)$$

where D_P - the soil water depletion
D_T - the total available water with in the root zone at field capacity.

The scheduler may use up to three soil horizons of different depths and water holding capacities to calculate the total available water in the root zone. The rooting depth is at its minimum from planting date to root development date. Rooting depth then increases from root development date to effective cover date. The maximum rooting depth is a function of crop but can be limited by compaction or texture of the soil profile.

---

(1) Reference to brand names is for the readers benefit and does not imply endorsement by the U.S. Government.

Immediately following an irrigation or rainfall, the water use will be near the reference ET because of evaporation from the soil surface. The additional water use beyond normal crop use because of evaporation on a given day is determined by the equation:

$$E_{tw} = K_w (0.9 - K_c) E_{tr} \qquad (3)$$

where K_w - a coefficient taking on the values of .8, .5, .3 for the first, second and third days respectively following rainfall or irrigation.

K_c - crop coefficient.

If K_c is greater than 0.9 or if no rainfall has occurred within the three previous days, then E_{tw} is zero. An exception is made for sandy soils (less than 0.1 mm/mm water holding capacity) so that E_{tw} is always zero on the second and third days following irrigation or rain.

Overview of SP1 operations

Three groups of data files are used with SP1. The E_t file stores the daily weather data from the area or "regional" weather station. The field data file stores field-specific data from a group of individual fields. For example: field name, crop, soil depth, water holding capacity, etc. The rain and irrigation file stores the date and amount of all the rain and irrigation events for each field in the field file.

At the beginning of the irrigation season the fields are grouped into field files. Disk data storage space must be allocated to store data for E_t files and field files. Next, the initial details of the E_t data files and the individual field files are entered using the editing functions of SP1.

The primary operations of SP1 are: to calculate E_t from recorded climatic data as well as to use calculated crop E_t, rainfall and irrigation amounts to calculate daily depletions in a field and thereby schedule future irrigations. The field soil moisture depletion may be adjusted to coincide with field observations immediately prior to scheduling or at any other time using SP1's editing functions. The field soil moisture depletions from the previous irrigation schedule are retained in each individual field file. If an error is found in a schedule it can, therefore, easily be corrected and rerun.

The scheduling strategy used by SP1 is as follows: E_t is calculated from the climatic data entered into the computer. Beyond this time E_t is predicted by SP1. Predicted E_t data are used to forecast daily soil moisture depletions.

The formula used to predict E_t for the forecast period is

$$E_t = 7.25 \exp -((D - 180)/100)^2 \qquad (4)$$

where D - the julian day number.

The above formula was calibrated to predict average E_t at Greeley Colorado. This equation would have to be re-calibrated for other sites.

Irrigations are scheduled so that the soil water depletion in the field does not exceed the maximum allowable depletion. When a time difference of more than 24 hours exists between irrigation at the start position and irrigation at the stop position in a field, the depletions are considered at both of these positions. Irrigations are scheduled so that both positions are maintained with less than the allowable depletion. This may mean that irrigation must commence before the start position has reached maximum allowable depletion in order that the stop position gets irrigated before it reaches maximum allowable depletion. The above strategy determines the most delayed date that irrigation can commence. Where the average application depth is less than the allowable depletion, then the field may be irrigated earlier without waste of irrigation water due to deep percolation beyond the root zone. The earliest date of irrigation, therefore, can be when the depletion equals the average application depth. This allows SP1 to indicate a range of days on which the farmer may irrigate his field. For example: Not earlier than June 1 and not later than June 10.

The computer printout schedule has been modified to better suit the field irrigation system. Each field record specifies the type of irrigation system in the field, i.e. center pivot, surface or regional. ('Regional' implies the estimates are not related to a specific field but are general estimates for a region.) This determines the form of the output. If a file is specified as "regional" then reference E_t and actual E_t for that crop will be printed. If a field record specifies 'center pivot' or 'surface' then a printed output specially designed for these irrigation systems is produced.

Field data recording procedures

Daily field data required for E_t estimates are as follows: maximum and minimum temperatures; average daily dew point temperature; solar radiation and wind run. Average dew point temperature is estimated by six simultaneous temperature-relative humidity readings spread throughout the day. Any number of temperature-relative humidity readings may be entered for each day or average dew point may be entered instead of the temperature-relative humidity readings.

Temperature recordings may by entered in either Celcius or Fahrenheit but all readings must be consistent. Solar radiation may be entered in Langleys, Watt-hours per sq meter or megajoules per sq meter. Wind run may be entered in miles or kilometers.

Rainfall and irrigation amounts may be entered in any measurement units. The user must provide a rainfall conversion factor and an irrigation conversion factor to convert the recordings to millimeters depth. These conversion factors are stored in the computer field file. This is to avoid any need for the irrigator to make any data conversions since this is a likely source of error. This procedure allows field readings of time (at a known flow rate) or volume (over a known area) to be used as irrigation depths. Rainfall can also be input as volume where the conversion factor for the cross sectional area of the gage has been stored.

Field data are recorded on specially prepared forms to enable direct entry into the computer. The evapotranspiration data field sheet is produced by normal printing methods but the rainfall field sheet and the irrigation field sheet are prepared by SP1. The rainfall field sheet lists all the rain gauges used by the fields in a field file. The irrigation field sheet lists all the fields in the field file by name and provides spaces to enter irrigation depths for any date and day of the week.

Data entry procedures and checks

Data is entered interactively with the computer. When SP1 is set in the data entry mode it prompts the user to enter the data in the correct order and performs a number of data checks as the data are being entered.

If an error is detected by the checking procedures an audible tone is generated by the computer and an appropriate prompt is shown on the screen. Where data is out of the normal range but still possibly correct the operator has the choice of correcting the entry or ignoring the prompt if the data is, in fact, correct. A clearly incorrect entry such as month greater than 12 will force the operator to re-enter a value which passes the error check.

Although SP1 uses Julian days for internal calculations this fact is transparent to the user. Dates are entered as digits in the form of 'month,day', e.g. 5,24. SP1 prints dates in the form 'MAY 24' and where appropriate will also output the day of the week, e.g. 'SUN MAY 24'. This is an attempt to avoid errors by working with days and dates which anyone can mentally cross check.

Program operation

To operate SP1, the program is first loaded into the computer processor. A list of options, called a 'menu', is then displayed on the computer screen from which the operator can select any one of eight options. The menu is the starting point for all these operations which, when completed, will return control to the program menu. At any time during the operation of the program, the operator may abort the present operation and return to the menu. Where it is necessary, SP1 asks the operator to identify the field file he wishes to work with. The menu options allow the operator to; allocate file space; enter and/or edit climatic, field and irrigation data; generate regional E_t using stored climatic data; and schedule field irrigations.

SP1 irrigation schedule printout

The scheduling procedure has three main stages shown on the computer printout. Each field record is coded so that printout is optional for each stage.

The first stage is the depletion update (Figure 1). This procedure accounts for rain, irrigation and E_t. It updates the soil depletions at the start and stop positions from the depletion update date shown on the field record to the most recent day of data in the E_t data file.

The second stage is the forecast depletion table (Figure 2). This table indicates the earliest date of irrigation and the most delayed date of irrigation for both the start and stop positions. The "*" indicates where the allowable depletion has been exceeded. Irrigation water may be applied at any time during the period shown by the "-" sign. The forecast table provides all the information necessary to schedule the next irrigation but it requires some interpretation by the user.

The third stage of the output gives a specific schedule for the particular irrigation system used on the individual field. Examples are shown in Figures 3 and 4. The present options are for center pivot and for surface irrigation systems.

The center pivot option (Figure 3) gives "not earlier than" and "not later than" starting dates with a range of rainfall allowances since the last irrigation. It also schedules subsequent irrigations to the end of forecast period.

The surface irrigation option (Figure 4) has a very abbreviated output. It uses only the date when start position reaches allowable depletion to predict the next irrigation date. Should the farmer require additional information he can opt to have the depletion update and/or the forecast depletion table printed also. The surface irrigation output provides a prediction of the

OWNER - SMITH REGION - CROOK
DATE - JUL 29 1981

FIELD - C-3 CROP - CORN

---------DEPLETION UPDATE----------

DAY	START WATER mm.	POSN USED in.	IRRIGATION AND RAINS mm.	in.	APPLICATION	.....DEPLETIONS..... START mm.	in.	STOP mm.	in.
THU JUL 23	6.0	0.24				6.0	0.24	10.2	0.40
FRI JUL 24	5.9	0.23	17.8	0.70	IRRIG	0.0	0.00	0.0	0.00
FRI JUL 24					EXCESS	005.9		005.7	
SAT JUL 25	5.6	0.22	15.2	0.60	RAIN	0.0	0.00	0.0	0.00
SAT JUL 25					EXCESS	009.6		009.6	
SUN JUL 26	1.2	0.05				1.2	0.05	0.0	0.00
SUN JUL 26					EXCESS			016.6	
MON JUL 27	5.6	0.22				6.8	0.27	5.6	0.22
TUE JUL 28	5.0	0.20				11.8	0.47	10.6	0.42
WED JUL 29	7.6	0.30				19.5	0.77	18.3	0.72

TOTAL EXCESS APPLICATION THIS PERIOD = 15.6 mm
NOTE: '*' INDICATES THE DEPLETION IS GREATER THAN ALLOWABLE

FIELD FILE - 'CR081' # 1 Et DATA - 'ETCR081' RAIN & IRRIG FILE - 'CR08101'
PLANTING DATE SUN MAY 10 EFFECTIVE COVER DATE TUE JUL 21
DEGREE DAYS SINCE MAY 21 = 1419 (FAHRENHEIT-DAYS BASE 50 METHOD)
CROP COEFFICIENT = .98 STRESS COEFFICIENT = .98

Figure 1. Typical depletion update.

FIELD - C-3 CROP - CORN

----FORECAST DEPLETION TABLE----

THE MAXIMUM USEFUL IRRIGATION AMOUNTS ARE SHOWN BELOW.
TIME TO IRRIGATE FIELD = 60 HOURS (2 DAYS)

DAY	START POSITN mm.		in.	STOP POSITN mm.		in.
THU JUL 30	26.0	-	1.02	24.8	-	0.97
FRI JUL 31	32.5	-	1.28	31.2	-	1.23
SAT AUG 1	38.9	-	1.53	37.7	-	1.48
SUN AUG 2	45.3	-	1.79	44.1	-	1.74
MON AUG 3	51.7	-	2.04	50.5	-	1.99
TUE AUG 4	58.1	-	2.29	56.9	-	2.24
WED AUG 5	64.4	-	2.54	63.2	-	2.49
THU AUG 6	70.7	-	2.78	69.5	-	2.73
FRI AUG 7	76.9	*	3.03	75.7	*	2.98
SAT AUG 8	83.0	*	3.27	81.8	*	3.22

AUG 8 ALLOWABLE DEPLETION = 73.2 mm
NOTE: '-' INDICATES DEPLETION > AVE APPLICATION DEPTH
NOTE: '*' INDICATES THE ALLOWABLE DEPLETION EXCEEDED

Figure 2. Typical forecast depletion table.

FIELD - C-3 CENTER PIVOT SYSTEM

IF 20.3 mm (.8 INCHES) ARE APPLIED IN 60 HOURS, THE STARTING TIMES ARE:

AMOUNT OF RAIN mm.	in.	NO SOONER THAN	NO LATER THAN	MAX USEFUL IRRIG DEPTH STARTmm	STOPmm
0.0	0.00	THU JUL 30	WED AUG 5		
5.0	0.20	THU JUL 30	THU AUG 6		
10.0	0.39	FRI JUL 31	FRI AUG 7		
15.0	0.59	SAT AUG 1	SAT AUG 8		
20.0	0.79	SUN AUG 2	SAT AUG 8		

ASSUMING THE SYSTEM WAS STARTED ON THU JUL 30 THE NEXT STARTING TIMES ARE:

AMOUNT OF RAIN mm.	in.	NO SOONER THAN	NO LATER THAN	MAX USEFUL IRRIG DEPTH STARTmm	STOPmm
0.0	0.00	SUN AUG 2	SAT AUG 8		
5.0	0.20	MON AUG 3	SUN AUG 9		
10.0	0.39	MON AUG 3	MON AUG 10		
15.0	0.59	TUE AUG 4	TUE AUG 11		
20.0	0.79	WED AUG 5	WED AUG 12		

ASSUMING THE SYSTEM WAS STARTED ON SUN AUG 2 THE NEXT STARTING TIMES ARE:

AMOUNT OF RAIN mm.	in.	NO SOONER THAN	NO LATER THAN	MAX USEFUL IRRIG DEPTH STARTmm	STOPmm
0.0	0.00	WED AUG 5	WED AUG 12		
5.0	0.20	THU AUG 6	THU AUG 13		
10.0	0.39	FRI AUG 7	FRI AUG 14		
15.0	0.59	FRI AUG 7	SAT AUG 15		
20.0	0.79	SAT AUG 8	SUN AUG 16		

Figure 3. Typical center pivot schedule.

OWNER - C.S.U. EXT SVCE REGION - GREELEY
DATE - AUG 2 1981

FIELD	CROP	LAST IRRIG ---DATE---	NEXT IRRIG ---DATE---	NEXT IRRIG 25 mm RAIN	AREA HA	FLOW M3/H	TIME HOURS	IRR DEPTH -mm-	--IN-	VOLUME CU. M	EFFY
CORN 1 SURGE	CORN	WED JUL 22	MON AUG 10	FRI AUG 14	2.3	185	16.1	124	4.9	2971	0.40
CORN 2 GP/R	CORN	FRI JUL 17	MON AUG 10	FRI AUG 14	2.6	185	17.7	124	4.9	3269	0.40
CORN 3 GP/R	CORN	SUN JUL 19	MON AUG 10	FRI AUG 14	4.2	185	28.8	124	4.9	5332	0.40
CORN 4A D/S PIT	CORN	SUN AUG 2	MON AUG 10	FRI AUG 14	2.1	185	6.3	55	2.2	1160	0.90
POTO 4B D/S	POTATOES	SUN AUG 2	SAT AUG 8	FRI AUG 14	1.0	185	4.0	70	2.8	736	0.40

Figure 4. Typical surface irrigation schedule.

REFERENCE Et (mm) FOR GREELEY 1981

	WED	THU	FRI	SAT	SUN	MON	TUE
JUL 1	7.9	4.9	5.3	5.9	7.1	7.2	7.6
JUL 8	6.0	6.7	6.7	7.3	6.0	5.1	7.2
JUL 15	6.7	3.7	5.2	4.8	6.8	6.4	6.9
JUL 22	4.5	7.0	6.9	5.4	2.5	4.7	5.1

ACTUAL Et (mm) CORN GREELEY 1981

	WED	THU	FRI	SAT	SUN	MON	TUE
JUL 1	5.3	3.4	3.8	4.3	5.2	5.5	5.9
JUL 8	4.7	5.4	5.5	6.0	5.0	4.4	6.2
JUL 15	5.9	3.3	4.6	4.3	6.2	6.0	6.4
JUL 22	4.3	6.7	6.5	5.2	2.3	4.4	4.8

Figure 5. Typical regional ET printout.

volume of irrigation water required and the time required at a given flow rate to apply that given volume. These predictions of time and volume are designed to correct the existing irrigation deficit at any specified efficiency. An option is available in the field record to have this surface schedule printed in metric or English units.

If any field file is specified as "regional" then the above three phases of output are replaced by a different form of output which tabulates the daily reference evapotranspiration for any specified period within the E_t data file. It also tabulates actual evapotranspiration by the crop specified on the field record without any allowance for soil moisture stress or surface evaporation following rain or irrigation (See Figure 5).

Further documentation on this program is available from the authors.

REFERENCES

Kincaid, D. C. and D. F. Heermann. 1974. Scheduling irrigations using a programmable calculator. U.S.D.A. Bulletin ARS-NC-12, 55 pp.

Heermann, D. F., H. R. Haise, R. H. Mickelson. 1976. Scheduling center pivot sprinkler irrigation systems for corn production in eastern Colorado. Trans. of the ASAE 19(2):284-287, 293.

Jensen, M. E. 1969. Scheduling irrigations using computers. J. Soil and Water Conserv. 24(5):193-195.

Jensen, M. E., J. L. Wright, B. J. Pratt. 1971. Estimating soil moisture depletion from climate, crop and soil data. Trans of the ASAE 14(5):954-959.

Jensen, M. E., D. C. N. Robb and C. E. Franzoy. 1970. Scheduling irrigations using climate-crop-soil data. Amer. Soc. Civil Engr., J. Irrig. and Drain. Div. 96(IR1):25-38.

Penman, H. L. 1963. Vegetation and Hydrology Tech Communication No. 53, Commonwealth Bur. of Soils, Harpenden, England, 125pp.

THE WATER MANAGEMENT AND CONSERVATION PROGRAM IN YOLO COUNTY, CALIFORNIA

Gordon Lyford
Member SCSA

Neil Schild
Member ASAE

The Bureau of Reclamation has been concerned with the efficient use of irrigation water for many years; for example, a series of "Use of Water Studies" was compiled from 1917 through 1960 on Federal Irrigation Projects (Debler, 1930; USBR, 1952 and 1962). These reports revealed the average onfarm irrigation efficiency to be 44 percent. Therefore, a new "Use of Water Study" was initiated in 1963 to "document accurately the present use of water on existing projects," and "to obtain information to improve and develop procedures to insure successful and efficient planning, designing, and operating of future projects" (USBR, 1978).

This new "Use of Water Study" compiled information on water deliveries, rainfall, consumptive use, surface runoff, solar radiation, daily temperatures, irrigation practices, cultural practices, crop yields, soils, and irrigation efficiencies. Irrigation efficiency refers to water stored in the root zone divided by water delivered to the field. These measurements were collected during the 9-year study at 17 locations throughout the seven Reclamation regions, and included 34 crops on 9,027 hectares and 970 fields. Results showed that the average seasonal irrigation efficiency was 48 percent, and could be improved to 60 percent efficiency by improved irrigation management.

Based on these results, and recognizing the benefits of improved irrigation, Reclamation began to develop its Irrigation Management Service (IMS) program in 1968. This program is a water balance method based on the evapotranspiration (ET) model developed by Dr. Jensen. This method relies on computing crop ET to determine soil moisture depletions from local weather conditions, thus minimizing laborious soil sampling procedures (Jensen, et al., 1970 and 1971). The goal of Reclamation's IMS program is, with the aid of a computerized technique, to efficiently schedule irrigation applications on individual fields and water deliveries throughout project conveyance and distribution systems.

In 1981, the IMS program was renamed the Water Management and Conservation (WMC) program to reflect a program of broader scope. The WMC program includes the IMS concepts as well as other aspects of operation, maintenance, and rehabilitation of Reclamation projects. The WMC program introduces districts to the use of minicomputers for water management, field scheduling, and lateral management as well as numerous recordkeeping and engineering functions. The computer program "IMS80" is the newest version of Reclamation's irrigation scheduling computer programs and is the heart of the WMC program. The "IMS80" program is presently operational on minicomputers in three districts and will soon be operational in other districts.

Gordon Lyford, Agricultural Engineer, Mid-Pacific Region, USBR, Sacramento.
Neil Schild, Assistant Regional Director, Mid-Pacific Region, USBR.

The "IMS80" program is designed to operate on the "lateral management" and "field scheduling" levels. The information required for field scheduling includes daily weather data, crop types, planting dates, field sizes, actual irrigation dates and quantities of water applied, and periodic measured soil moisture depletions. The participating water users are then provided weekly computer reports advising them of the proper timing and quantity of the next irrigation on each field. Lateral management utilizes all of the information required for field scheduling except for the measured soil moisture depletions. The results of lateral management provide a district with information on daily lateral diversion requirements, system efficiencies, field irrigation efficiencies, and improved water accounting and management capabilities.

In 1980, there were 19 irrigation districts with WMC programs which involved 63,941 hectares in 2,130 fields and on 500 farms. This included both field scheduling and lateral management efforts (USBR, 1980). The following describes the operation of the WMC program in the Yolo County Flood Control and Water Conservation District (District).

Yolo County Flood Control and Water Conservation District

The District is located in the southeast portion of Yolo County northwest of Davis, California. The District covers about 22,000 hectares where a wide variety of crops is grown, including tomatoes, beets, corn, alfalfa, orchards, rice, wheat, sunflowers, and melons. The most important of these crops is tomatoes, as Yolo County is one of the largest tomato-growing regions in the world.

The water supplies for the District include pumped underground water, Clear Lake, and Indian Valley Reservoir. The latter was constructed with the assistance of a loan from Reclamation's Small Loan Program. The water distribution system consists mostly of unlined canals; and the area's first irrigation canal, the Moore Canal, was constructed in 1864 to divert water from Cache Creek to lands south of Woodland (California, 1979).

The WMC program was introduced to the District in 1980 with a cooperative agreement between the District and Reclamation that will end in December 1982. During the agreement, Reclamation will convert the "IMS80" computer program package to run on the District's computer, and the District will provide employees to learn the techniques of the WMC program. By the end of the 3-year program, the District will be able to run the WMC program on their own if they elect to do so.

With the current focus of computer technology on smaller computers, most districts should be able to justify purchase of a minicomputer to perform all of their functions for less than $20,000. The District's computer was purchased in 1975 and includes a 32-k processor, two 256-k flexible disk drives, a 200 cps matrix printer, and a 64-character-wide CRT. The "IMS80" program package was converted to Basic and made operational in March 1980, prior to the irrigation season.

IMS80

The "IMS80" computer package was designed and written in January 1980 in the FORTRAN language. The intent of "IMS80" was a simpler program based on Reclamation's experience since 1968 in the numerous versions of irrigation scheduling computer programs. "IMS80" is easy to convert to minicomputers, and incorporates the lateral management logic in a less complicated form than in previous programs. The "IMS80" program listings and documentation are available from Reclamation by writing to the Bureau of Reclamation, Engineering and Research Center, Technical Publications Branch, Denver

Federal Center, Denver, Colorado 80225. The "IMS80" computer program includes three data files and 14 programs. The three data files are the "Weather" file, "Crop" file, and the "Field" file, and the various programs manage these data files to produce a variety of outputs. An example of the farmer's weekly report is shown in figure 1.

A weather station was established in March 1980 next to a golf course in a representative area of the district to gather the necessary climatic measurements. The station includes a maximum-minimum thermometer, wet and dry bulb thermometers, totalizing anemometer, class A evaporation pan, rain gauge, and a solar meter. The station is read each day at eight in the morning during the irrigation season by District employees. These weather readings are used in the computer program to calculate daily evapotranspiration potentials (ETP) by both the Jensen and Penman methods. The Penman equation used in the "IMS80" Basic program is adapted from the United Nations Irrigation and Drainage Paper Number 24. The Jensen equation is being used with the constants CT and TX set to 0.012 and 17.0, respectively (Jensen, 1973). All of these weather readings and ETP calculations are stored in the "Weather" data file.

The "Crop" file contains the calculated historic and forecast daily evapotranspiration (ET) values for each crop. These ET values are used to calculate the soil moisture depletions and to forecast irrigation dates for each field being monitored. The "Crop" file also contains the planting, cover, and termination dates and current crop coefficient (KC) for each crop. The daily ET values for each crop are calculated with the equation ET=KC x ETP.

The "Field" file contains specific information for each field such as the past irrigation date and amount, the crop type, the measured soil moisture depletion, the allowable depletion, and the forecast irrigation date and amount. The "Field" file is used as the basis of lateral management programs since it also contains information on water deliveries, crop acreage, turnout identifications, turnout capacities, and farm identifications.

Field Procedure

Field work and computer updates must be coordinated to minimize the computer time, optimize time in the field, and provide farmers with timely reports each week. Each field man has his own area of responsibility and visits about fifteen fields each day. In 1981, computer updates were performed Tuesday and Thursday afternoons for a total of 3 hours of computer time each week. With this procedure, each scheduled field will be checked at least once each week and each farmer will receive his irrigation report the same day of each week.

Field checks are required on a regular basis in order to verify the computer model. The information gathered during field visits may include irrigation dates, planting dates, crop development, soil moisture depletions, soil conditions, rooting depth, and perhaps pest problems. It is important to contact the farmer periodically to inform him of any problems and to personally discuss and review the soil moisture status of each field. The technique that we are using to measure soil moisture depletions is the "field" method (Merriam, 1960).

Early in the season it is important to check the planting depth of the seeds in relation to the thickness of any soil crusting. Oftentimes growers exaggerate the crusting problems and apply excess irrigations early. Near the end of the season, the soil moisture storage should be checked in relation to the cumulative crop ET in order to time the last irrigation. On corn fields it is typical for farmers to apply one or more extra irrigations at the end of the season, thus failing to utilize the available moisture stored in the root zone. In many cases it is worth the time to sample the soil to determine the moisture status of a field before planting and after harvest. This can help determine pre-irrigation requirements, or help verify a crop ET model.

During field visits each field man uses a worksheet, similar to the example in figure 2, to record notes for use in computer updates. The sector numbers on the worksheets are used for direct access memory changes in the "Field" data file. During field visits the measured depletion by the "field" method should be compared to the computed depletion on the worksheet. If a large difference is found between the computed depletion and measured depletion, more samples should be taken until a representative value is determined. Any other changes in field data, such as irrigation events, allowable depletions, or water orders, should also be noted on the worksheet and entered into the "Field" file during the computer update.

Program Size

The following table summarizes the acreage under the WMC program in the District in 1980 and 1981.

	1980		1981	
Crop	Hectares	Fields	Hectares	Fields
Tomatoes	537	20	588	19
Corn	442	11	688	23
Alfalfa	367	17	53	2
Orchards	156	7	124	4
Beets	198	7	210	6
Sunflowers	61	1	30	1
Wheat	61	1	91	2
Melons	53	1	8	1
Total	1,875	65	1,792	58

The total farms on the WMC program included 15 in both 1980 and 1981. The major changes from 1980 to 1981 were fewer alfalfa fields and more corn fields. The farmers felt that it was difficult following the irrigation schedules on alfalfa because the cutting dates and irrigation dates often conflicted. However, many farmers were able to save two irrigations on corn by following the irrigation schedules in 1980; therefore, they put more corn fields on the program in 1981. The District charged the farmers a fee of $2.47 per hectare for the scheduling service each year.

Results

The benefits of improved irrigation scheduling are very difficult to document due to the numerous factors that determine crop yields. The difficulties in comparing crop yields in a field situation between scheduled and nonscheduled fields relates to the lack of a control and the degree to which the irrigation schedule is followed. We know, in general, fewer irrigations were applied to the scheduled corn and tomato fields than on nonscheduled fields based on conversations with farmers.

In order to estimate the results of the WMC program in 1980, the yield of tomatoes can be examined. The average tomato yield for Yolo County in 1980 as reported by the California Crop and Livestock Reporting Service was 53.6 Mg/hectare. The average tomato yield on the scheduled fields in the District was 56.7 Mg/hectare in 1980. The average tomato yield on the fifteen fields that followed the recommendations closest was 61.9 Mg/hectare. A ton of tomatoes sells for about $50; therefore, a relatively small yield increase is important. We feel that these results are positive when one considers that the fields on the scheduling program were some of the most difficult fields to irrigate. This is because the farmers request help in areas where they feel a less than desirable result is being achieved.

It is also difficult to evaluate if the quantity of applied water has been improved on the scheduled fields. This is because ground-water applications are often not measured, and measured surface water is often mixed with ground water and split to several fields in unknown ratios. We do know that applications have been reduced on some scheduled fields, and many orchard crops are underirrigated due to inadequate sprinkler systems.

On many crops it is possible to observe improved irrigation practices on scheduled fields by comparison to local "rule of thumb" practices on nonscheduled fields. In this area it is common practice to irrigate tomatoes and sugar beets two times just to germinate the seeds and establish a good stand. We have demonstrated to some farmers that a good tomato or sugar beet stand can be achieved with one irrigation. This represents a savings of about 15 cm of water and $30/hectare including labor, water, and pumping costs. With corn we have shown that one or two irrigations can be saved during the season compared to normal practices.

Conclusions

Water management by districts and farmers can be improved, and crop yields can be increased by using an information system such as Reclamation's WMC program. Water districts should put more effort into providing farmers with timely information on the quantity of each irrigation application. Districts should also use WMC information to better manage distribution system operations, such as determining more accurate dates for distribution system startup and shutoff. The key to improved water management is to make use of real time information for district and farm operations. Reclamation's WMC program can provide the tools to accomplish these goals.

FIGURE 1

THIRSTY ACRES
#1 MAIN STREET
ANY TOWN
CA 12345

FARM # 645

FIELD IRRIGATION SCHEDULE AS OF: JUN 18

LATERAL	TURNOUT	FIELD NUMBER	ACRES	CROP	LAST IRRIGATION DATE	LAST IRRIGATION AMOUNT INCHES	FORECAST ET RATE IN./DAY	ALLOWABLE DEPLETION INCHES	NEXT IRRIGATION DATE	NEXT IRRIGATION AMOUNT INCHES	CFS	(AC X IN)	HRS
WIN	1336R	2	20	MELONS	JUN 14	7.5	0.31	4.0	JUN 26	5.3		106.6	
BUK	060L	1	75	BEETS	JUN 11	5.8	0.20	3.0	JUN 28	3.9		299.9	

FIGURE 2

IRRIGATION SCHEDULING WORKSHEET AS OF 618

FILE = BFIELD
SECTOR = 49

LATERAL: WIN TURNOUT: 1336R FARM # 645

DATE VISITED:

FIELD	ACRES	CROP	ID	KC	ET ADJ %	IRR EFF %	AVG DAY USE	SUM ET TOD	DEPLETIONS ALW IN.	DEPLETIONS TOD IN.	DEPLETIONS MDP IN.	DEPLETIONS MDP DATE	IRRIGATIONS SUM	IRRIGATIONS LAST DATE	IRRIGATIONS AMT IN.	IRRIGATIONS NEXT DATE	IRRIGATIONS AMT IN.
*****	*****	**********	**	****	***	***	***	****	****	****	********		***	****	***	****	***
2	20.	MELONS	10	0.81	0	75	0.31	7.1	4.0	1.36	2.50	608	2	614	7.5	626	5.3

IRRIGATION SCHEDULING WORKSHEET AS OF 618

FILE = BFIELD
SECTOR = 63

LATERAL: BUK TURNOUT: 060L FARM # 645

DATE VISITED:

FIELD	ACRES	CROP	ID	KC	ET ADJ %	IRR EFF %	AVG DAY USE	SUM ET TOD	DEPLETIONS ALW IN.	DEPLETIONS TOD IN.	DEPLETIONS MDP IN.	DEPLETIONS MDP DATE	IRRIGATIONS SUM	IRRIGATIONS LAST DATE	IRRIGATIONS AMT IN.	IRRIGATIONS NEXT DATE	IRRIGATIONS AMT IN.
*****	*****	**********	**	****	***	***	***	****	****	****	********		***	****	***	****	***
1	75.	BEETS	32	0.52	0	75	0.20	3.7	3.0	0.81	0.20	615	2	611	5.8	628	3.9

1. California, State of, 1979. The California Water Atlas. General Services, North Highlands, CA, 118 p.

2. Debler, E. B., 1930. Use of the Water on Federal Irrigation Projects. Trans. of the ASCE: 1195-1241.

3. Jensen, M. E., D. C. N. Robb and C. E. Franzoy, 1970. Scheduling Irrigations Using Climate-Crop-Soil Data. J. Irrig. and Drain Div., ASCE 96: 25-28.

4. Jensen, M. E., J. L. Wright and B. J. Pratt, 1971. Estimating Soil Moisture Depletion from Climate, Crop and Soil Data. Trans. of the ASAE 14: 954-959.

5. Jensen, M. E., 1973. Consumptive Use of Water and Irrigation Water Requirements. American Society of Civil Engineers, New York, NY, 215 p.

6. Merrian, J. L., 1960. Field Method of Approximating Soil Moisture (Deficiency) for Irrigation. Trans. of the ASAE, Vol 3, No. 1.

7. U.S. Bureau of Reclamation, 1952. Use of Water on Federal Irrigation Projects. USBR, Denver, CO.

8. U.S. Bureau of Reclamation, 1962. Use of Irrigation Water, A Report on Selected Federal Reclamation Projects, 1949-1960. USBR, Denver, CO, 116 p.

9. U.S. Bureau of Reclamation, 1978. Use of Water on Federal Irrigation Projects, Summary of 1964-1972 Studies. USBR, Denver, CO, 87 p.

INTERFACING IRRIGATION ELECTRIC LOAD MANAGEMENT AND IRRIGATION SCHEDULING

L. E. STETSON

ABSTRACT

Several rural electric power suppliers in Nebraska have been controlling irrigation pump motors, based on peak electrical demands. Experimental control started in 1973. In 1975, 1976, and 1977, more comprehensive control of irrigation pump motors was initiated. Peak power demands have been reduced without reducing crop yields.

Because of the success of these programs in controlling peak power demands, the concept of irrigation electrical load management is spreading throughout the United States. Those who provide irrigation scheduling information may need to consider control of electric power in predicting when to irrigate.

Data from the cooperating power suppliers who have had irrigation load management programs will be presented to illustrate examples of control options and actual control which resulted. Actual control of irrigation motors has not been as severe as the power contracts permit. For example in 1980, a very dry year in the Great Plains, motors under contract for any time control were shut off less than 90 hours during the entire irrigation season. Data and analysis will include at least five power suppliers and will encompass two to seven years of history. Control contracts include one day, two days, and any time control.

The author is: L. E. Stetson, Agricultural Engineer, USDA-ARS, Lincoln, NE.

IRRIGATION SCHEDULING BY THE CHECKBOOK METHOD

Darnell R. Lundstrom
Member ASAE

Earl C. Stegman
Associate Member ASAE

Hal D. Werner
Member ASAE

Determining when to irrigate and how much water to apply is a critical decision for an irrigator. Irrigation scheduling can be accomplished using direct reading methods such as tensiometers and soil moisture blocks as indicators of soil moisture. Several computational methods have also been developed which an irrigator may use to estimate crop water use and to calculate soil moisture contents (Jenson, et al., 1971; Black and Brosz, 1972; Irrigation Guide for Minnesota, 1976; Vitosh, 1977; Vitosh and Kidder, 1975).

Allowing soil moisture depletion to become excessively high between irrigations reduces crop transpiration and consequently reduces yield. Under North Dakota conditions several direct relationships have been shown between total evapotranspiration during the growing season and crop yield (Stegman and Bauer, 1977; Stegman and Lemert, 1981; Stegman and Nelson, 1973). Different crops vary in their response to root zone soil moisture content and a crops response also varies significantly with ambient temperature (Stegman, et al., 1976). Higher temperatures cause stress at lower soil moisture deficits.

The checkbook method was developed as a relatively simple method of on-farm scheduling. The method uses maximum air temperature, long-term average solar radiation and crop growth stages.

Development of the Checkbook Method

The checkbook method of irrigation scheduling (Lundstrom and Stegman, 1976) was developed in the mid 1970's during a period of rapid irrigation development in North Dakota and neighboring states. Inexperienced irrigators needed a method for estimating crop water use and a simplified technique for determining when to irrigate.

Crop coefficient curves were being developed (Stegman and Valer, 1972) for water balance methods of scheduling in anticipation of the availability of consultant services. However, the need for simplified techniques was also apparent because not all farmers would want or be able to obtain the more sophisticated management service. The checkbook system was developed utilizing available information from the Jensen et al. (1971) model. The Jensen model estimates water use as a function of daily crop coefficients, daily mean air temperature and daily solar radiation. The checkbook system utilizes crop coefficients, daily maximum air temperature as an estimator of daily mean temperature and solar radiation estimates that are based on long-term monthly averages.

Daily water use tables, dependent on maximum daily temperatures and stage of crop growth, were developed. After determining when his crop emerges, an

The authors are: Darnell R. Lundstrom, Extension Agricultural Engineer, North Dakota State University, Fargo, N.D.; Earl C. Stegman, Professor, Agricultural Engineering Dept., North Dakota State University, Fargo, N.D.; Hal D. Werner, Extension Irrigation Engineer, Area Extension Office, University of Minnesota, Staples, MN.

irrigator reads the estimated daily water use from a table using week after emergence and maximum daily air temperature. A water-balance (checkbook) sheet is kept to determine when to irrigate, avoiding excessive soil water depletions yet not irrigating excessively.

The checkbook scheduling method is a basic system for on-farm water management. The system is easy to understand and use while still maintaining an acceptable level of accuracy. The checkbook system does not have the sophistication of computer-based scheduling methods as are used by many irrigation consultants and it is not meant to replace these services. Rather it is an on-farm water management system.

The system was rapidly adopted by many irrigators in North Dakota and neighboring states, particularly Minnesota. An Extension bulletin was prepared for Minnesota irrigators (Werner, 1978) and the "Irrigators Association of Minnesota" has printed a pocket sized "Checkbook" (Werner, 1979), patterned after the Minnesota bulletin.

As with other methods of scheduling, field verification of the estimated soil moisture content is required. A soil probe to manually check moisture profiles is recommended. Soil moisture measurement devices such as tensiometers or moisture blocks may also be used.

Development from Jensen Scheduling Model

The Jensen et al. (1971) Scheduling Model has been calibrated (via locally developed crop coefficients) for irrigation scheduling in North Dakota. Crop coefficients have been developed for a number of crops (Stegman and Valer, 1972; Stegman et al., 1977). These coefficients were adapted to the checkbook system as follows.

The Jensen Model estimates daily crop evapotranspiration as:

$$ET = f[K_c, T_a, R_s]$$

where ET = estimated daily crop evapotranspiration
f = a function of
K_c = a crop coefficient adjusted for available soil moisture and increased evaporation from the soil surface occurring after rain or irrigation
T_a = daily mean air temperature
R_s = daily solar radiation

The first step in preparing the checkbook system was the development of a set of potential evapotranspiration tables, Table 1. These tables were developed from average monthly solar radiation data and five ranges of daily mean air temperatures based on North Dakota averages.

It was assumed irrigators would have more ready access to daily maximum than to daily mean tempertures so the table was shifted to reflect these values by adjusting the mean temperatures upward by 6.7 to 7.8°C (12 to 14°F) depending on month (Daily Normals and Record Extremes, 1976).

The checkbook method estimates daily potential evapotranspiration as:

$$ET = f[K_{co}, T_{max}, R_{s_{ave}}]$$

where: ET = estimated daily crop evapotranspiration
f = a function of
K_{co} = a weekly crop coefficient reflecting green ground cover
T_{max} = daily maximum air temperature
$R_{s_{ave}}$ = daily average solar radiation based on monthly averages

Table 1: Typical, daily crop water use amounts, as related to time of season and maximum daily air temperature in North Dakota.

	Maximum Daily Air Temperature ^{o}C (^{o}F)				
	10-15 (50-59)	15.6-20.6 (60-69)	21.1-26.1 (70-79)	26.7-31.7 (80-89)	32.2-37.2 (90-99)
Month	Potential Crop Water Use CM/Day (Inches/Day)				
May	.18 (.07)	.33 (.13)	.48 (.19)	.58 (.23)	.69 (.27)
June	.23 (.09)	.38 (.15)	.53 (.21)	.64 (.25)	.74 (.29)
July	.23 (.09)	.38 (.15)	.56 (.22)	.66 (.26)	.76 (.30)
August	.20 (.08)	.33 (.13)	.46 (.18)	.56 (.22)	.66 (.26)
September	.18 (.07)	.25 (.10)	.33 (.13)	.43 (.17)	.51 (.20)

Estimated crop water use tables have been prepared for several crops. These were prepared by applying the crop coefficient curves, as developed for the Jensen Model (Stegman, et al., 1977: Stegman and Valer, 1972), to the potential evapotranspiration table. These crop coefficient curves were modified slightly in that the weekly K_{co} values were not allowed to exceed 1.0. This adjustment helped compensate for effects of available water level in the root zone. The Jensen Model compensates for soil moisture effects with its K_c value, while the checkbook system does not. Average crop emergence dates were based on observations at experimental sites. Average water use for corn is shown in Table 2. Water use tables are also available for alfalfa, wheat, sugarbeets, pinto beans and sunflower.

Table 2: AVERAGE CORN WATER USE CM/Day (In/Day)

Week After Emergence	4	5	6	7	8	9	10	11	12	13	14	15	16	17
Date														
Temperature 10-15°C (50-59°F)	.10 (.04)	.13 (.05)	.18 (.07)	.20 (.08)	.23 (.09)	.23 (.09)	.23 (.09)	.23 (.09)	.23 (.09)	.23 (.09)	.20 (.08)	.20 (.08)	.15 (.06)	.13 (.05)
15.6-20.6^{o}C (60-69°F)	.18 (.07)	.23 (.09)	.28 (.11)	.33 (.13)	.38 (.15)	.38 (.15)	.38 (.15)	.36 (.14)	.36 (.14)	.36 (.14)	.33 (.13)	.33 (.13)	.23 (.09)	.18 (.07)
21.1-26.1^{o}C (70-79°F)	.25 (.10)	.30 (.12)	.38 (.15)	.48 (.19)	.56 (.22)	.56 (.22)	.56 (.22)	.51 (.20)	.51 (.20)	.51 (.20)	.46 (.18)	.43 (.17)	.28 (.11)	.23 (.09)
26.7-31.7^{o}C (80-89°F)	.30 (.12)	.38 (.15)	.46 (.18)	.58 (.23)	.64 (.25)	.66 (.26)	.66 (.26)	.61 (.24)	.61 (.24)	.61 (.24)	.56 (.22)	.53 (.21)	.36 (.14)	.30 (.12)
32.2-37.2^{o}C (90-99°F)	.33 (.13)	.43 (.17)	.53 (.21)	.66 (.26)	.74 (.29)	.76 (.30)	.76 (.30)	.74 (.29)	.74 (.29)	.74 (.29)	.66 (.26)	.64 (.25)	.43 (.17)	.36 (.14)

Procedure for Use by Irrigators

A detailed step-by-step procedure is provided to checkbook users. A soil moisture balance sheet, Table 3, is used to keep an account of soil moisture levels, irrigation and rainfall. The following procedure is used in North Dakota. Minnesota uses a similar method, but utilizes a soil moisture balance (positive balance) rather than a soil moisture deficit (minus balance) as used in North Dakota.

In Figure 3 the example procedures are:

1. Initial entries include the crop (corn), emergence date [May 20], and maximum root zone available moisture capacity [11.4 cm (4.5 inches)].

Root zone storage capacities are usually based on the coarser textured soils in a field, unless they represent only a small area.

Table 3: Example of Soil Moisture Balance Sheet Used by Irrigators with Checkbook Scheduling Method.

SOIL MOISTURE BALANCE SHEET -
CROP CORN MAX. AVAIL 11.4 CM (4.5 IN.) MOISTURE IN ROOT ZONE
EMERGENCE DATE MAY 20

Week	Date	Max Temp °C(°F)	Crop Water Use CM(in)	Irrigation CM(in)	Rain-Fall CM(in)	Soil Moist Deficit CM(in)
10	7/22	25 (77)	.56 (.22)			2.16 (.85)
	23	26 (78)	.56 (.22)		.58 (.23)	2.14 (.84)
	24	27 (80)	.66 (.26)			2.80 (1.10)
	25	28 (82)	.66 (.26)			3.46 (1.36)
	26	31 (88)	.66 (.26)	2.74 (1.08)		1.38 (.54)
	27	32 (90)	.76 (.30)	x x		2.14 (.84)
	28	33 (92)	.76 (.30)	x x		2.90 (1.14)
11	29	34 (94)	.74 (.29)	x		3.64 (1.43)
	30	34 (94)	.74 (.29)	2.74 (1.08)		1.64 (.64)
	31	29 (84)	.61 (.24)	x		2.25 (.88)
	8/1	24 (76)	.51 (.20)	x	.51 (.20)	2.25 (.88)
	2	26 (79)	.51 (.20)	x	2.13 (.84)	.63 (.24)
	3	28 (82)	.61 (.24)			1.24 (.48)
	4	27 (80)	.61 (.24)			1.85 (.72)
12	5	23 (73)	.51 (.20)		1.57 (.62)	.79 (.30)
	6	28 (82)	.61 (.24)			1.40 (.54)
	7	30 (86)	.61 (.24)			3.18 ~~2.01~~ ~~(.78)~~ 1.25
	8	26 (78)	.51 (.20)	2.74 (1.08)		.95 (.37)
	9	24 (76)	.51 (.20)	x		1.46 (.57)
	10	28 (82)	.61 (.24)	x		2.07 (.81)
	11	31 (88)	.61 (.24)	x		2.68 (1.05)

2. System pumping rate is converted to the net irrigation amount that is applied per irrigation, i.e. per field revolution for center pivot systems. In Table 3, the irrigation amounts reflect one field revolution per four day period with a net application equal to 2.74 cm (1.08 inches). With an expected water use rate of .64 cm (.25 inches) per day, the irrigator started water application when the soil water deficit approached 3.2-3.8 cm, (1.25 to 1.50 inches) thus avoiding a deficit larger than 50 percent of the root zone water holding capacity [5.75 cm (2.25 in.)] at the end of the irrigation.

3. Estimates of crop water use are read from the appropriate water use table. The user must know only the week after emergence and the maximum daily air temperature.

4. Daily crop water use is added to the soil moisture deficit. Irrigation water and rainfall are subtracted from the deficit. The deficit <u>cannot become less than zero</u> in cases of excess irrigation or rainfall.

5. Beginning soil moisture deficit is usually estimated by using the feel or appearance method and a hand probe. Detailed procedures for the hand probe are also provided with the checkbook system.

6. In the example, the field was checked or probed on August 7, in the 12th week after emergence and a determination was made that the moisture deficit was 3.18 cm (1.25 in.) rather than the 2.01 cm (.78 in.) as was estimated. Thus a correction was made. Users are advised to check fields on a weekly or biweekly interval. Probing is especially important to determine the real effect of several light rains or a heavy rain where the total rainfall may not have entered the soil profile.

7. Scheduling may start in any week during the irrigation season. The field is hand probed to establish an initial soil moisture deficit and soil water balance computations are begun using ET values for the correct week since crop emergence.

Comparison of Checkbook Method with Other Scheduling Methods

Figures 1 and 2 show the comparisons of cumulative ET for corn as estimated using the checkbook and the Jensen Scheduling Model. In both years the checkbook method paralleled the Jensen Model quite closely. The 1975 season was close to average with temperatures during the mid-April thru September growing season averaging 0.16°C (0.3°F) above normal. April was considerably below normal with May-June being above normal and July considerably above normal. August and September were slightly below normal. The 1976 season was a very warm year averaging 2.16°C (3.9°F) above normal for the mid-April thru September season. April, May, June, July, August and September were all considerably above normal. The 1976 climatic data served as a check on how the checkbook system will respond to a year with atypical temperatures. In 1976 at 120 days the checkbook system fell behind. At that time, the checkbook system shows the corn maturing. However in 1976 because of the warm temperatures and longer growing season, the corn was still actively growing and requiring water. This shows the necessity for the irrigator to observe his crop and make adjustments as necessary.

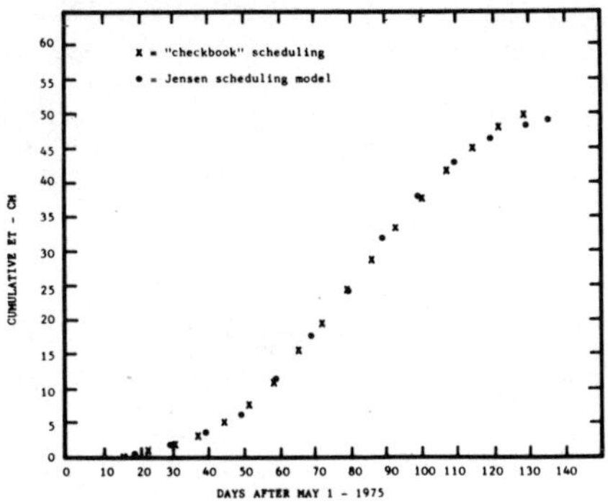

Figure 1. Comparative ET accumulations for irrigated corn - Oakes, North Dakota, 1975

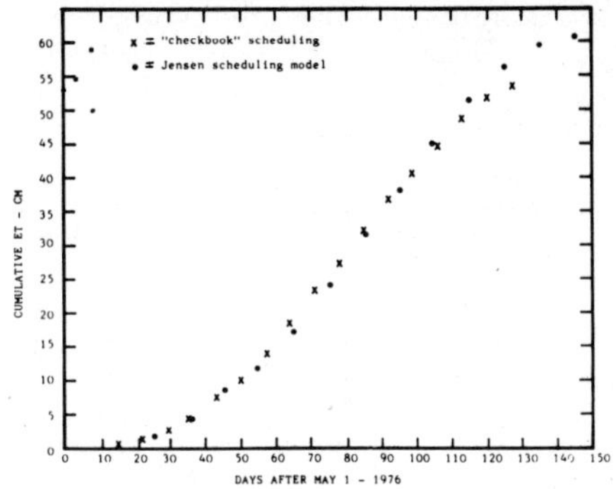

Figure 2. Comparative ET accumulations for irrigated corn - Oakes, North Dakota, 1976.

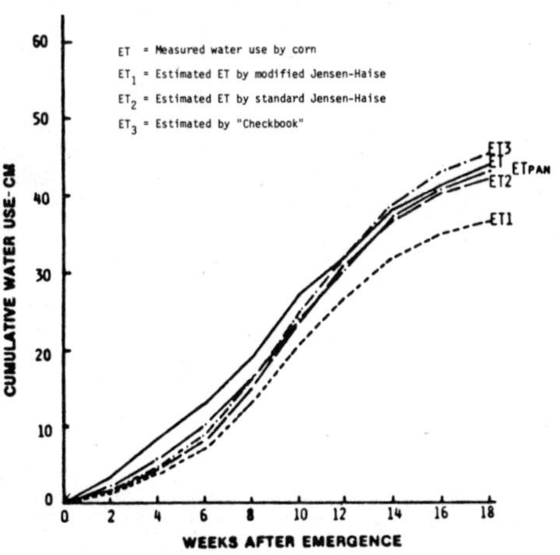

Figure 3. 1977 Cumulative Corn ET and Calculated Corn ET (from Dylla *et al*, 1980)

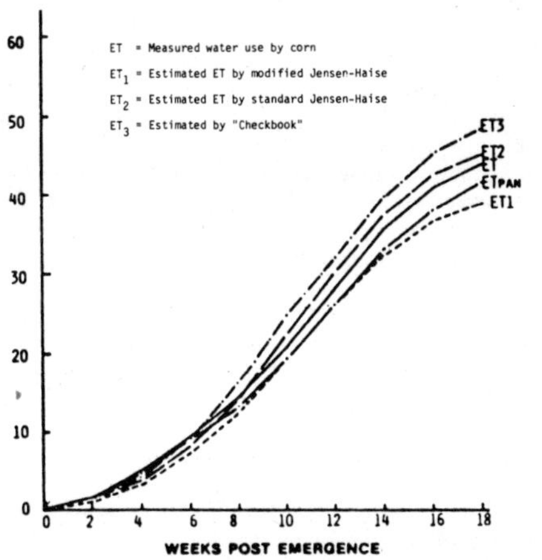

Figure 4. 1978 Cumulative Corn ET and Calculated Corn ET (from Dylla *et al*, 1980)

Dylla, et al. (1980) compared the measured ET of corn in west central Minnesota with estimated ET by: (a) the original or standard Jensen-Haise equation, (b) the modified Jensen-Haise equation, (c) the checkbook scheduling method and (d) U.S. Weather Bureau pan evaporation measurements with polynomial crop coefficients applied. The crop coefficient curve for (a), (b), and (c) was from Stegman et al. (1979). A comparison of these methods for the years 1977 and 1978 is shown in Figures 3 and 4. Dylla, et al. (1980), concluded that "The checkbook method, a simplified solution of the standard Jensen-Haise equation for southeastern North Dakota crops, requiring only daily maximum temperature and precipitation measurements and a set of crop water use tables to determine water used, was sufficiently accurate for most farmers and irrigators to use for determining when to irrigate".

At the Staples Irrigation site, Staples, Minnesota, the checkbook system was compared with tensiometers, evaporation pan with crop coefficient, .64 cm (.25 in.) per day irrigations and a check (no irrigation) (Begsrud, et al., 1980). Results with corn for the years 1977, 1978 and 1979 are shown in Table 4.

Table 4: Evaluation of different scheduling methods, on corn, 3 year average, 1977-1979, irrigation site, University of Minnesota, Staples.

Method	Yield Average Ton/Hectar (BU/Ac)	Average Net Water Applied CM (in)
.64 cm (.25 in)/day	8.81 (140)	41.1 (16.2)
Checkbook	9.06 (144)	26.2 (10.3)
Tensiometer	8.77 (139.5)	23.6 (9.3)
Evaporation Pan w/crop coeff.	8.60 (136.7)	22.4 (8.8)
Check	4.44 (70.6)	---

The checkbook system gave the highest yield however, it required slightly more irrigation water than the tensiometer or evaporation pan with crop coefficient methods.

Summary

The checkbook method provides irrigators with an on-farm water balance method for irrigation scheduling. The system was developed from work with the Jensen Scheduling model. The Jensen Scheduling model estimates water use as a function of daily mean air temperature, daily solar radiation and crop coefficient curves that reflect green ground cover effects. ET is additionally modified for effects of available soil water and wetness of the soil surface. The checkbook system utilizes locally developed crop coefficient curves, daily maximum temperature as an estimator of daily mean temperature and solar radiation based on long term monthly averages. The system reduces to daily water use tables which are dependent on maximum daily temperature and stage of crop growth.

To use the checkbook method an irrigator estimates daily water use from a crop specific water use table that is referenced to week after emergence and maximum daily temperature. By keeping a water-balance (checkbook) sheet he determines when to irrigate, avoiding excessive soil water depletions, yet not irrigating excessively.

The system has been used widely in North Dakota and Minnesota by practicing irrigators. Researchers at the Staples, Minnesota irrigation site have compared the checkbook system to other scheduling methods and found it equal to their other methods. Comparison of the checkbook system with the more sophisticated Jensen model showed good correlation in southeast North Dakota in both a close-to-average crop year (1975) and an abnormally hot-dry year (1976). Dylla, et al. (1980) found the system to be sufficiently accurate for most farmers and irrigators to use for determining when to irrigate.

The checkbook system meets the original objective, i.e., to provide a system which an irrigator can use to schedule his own irrigations with a reasonable level of accuracy. The system appears to respond to abnormal seasonal temperature variations. Although this system was developed using data form southeastern North Dakota it would be expected to deliver suitable accuracy under similar conditions of latitude, elevation and summer climate.

REFERENCES

1. Bergsrud, F., H. Werner, M. Wiens. 1980. Irrigation Scheduling, Staples, 1977-1980. Demonstration Report. Staples Irrigation Center, University of Minnesota.

2. Black S., D. Brosz. 1972. Irrigation Scheduling Through Moisture Accounting. Extension Circular 686. South Dakota State University, Brookings.

3. Daily Normals and Record Extremes of Temperature Based on 1941-1970 Data. 1976. Climatic Data Laboratory, Department of Soils, North Dakota State University, Fargo.

4. Dylla, A. S., D. R. Timmons, H. Schull. 1980. Estimating Water Use by Irrigated Corn in West Central Minnesota. Soil Science Society American Journal. 44:823-827.

5. Irrigation Guide for Minnesota. 1976. Soil Conservation Service, USDA, St. Paul, Minnesota. Pp 7-1 to 7-17.

6. Jensen, M. E., J. L. Wright 2nd, B. J. Pratt. 1971. Estimating Soil Moisture Depletion from Climate, Crop and Soil Data. Trans of ASAE 14(5):954-959.

7. Lundstrom, D. R., E. C. Stegman. 1976. Irrigation Scheduling by the Checkbook Method. Extension Agricultural Engineering Dept., North Dakota State University, Fargo.

8. Lundstrom, D. R., and E. C. Stegman. 1977. Checkbook Method of Irrigation Scheduling. ASAE Paper NCR 77-1001, ASAE, St. Joseph, Michigan.

9. Stegman, E. C., A. Bauer. 1977. Sugarbeet Response to Water Stress in Sandy Soils. Transactions from ASAE 20(3): 469-473, 477.

10. Stegman, E. C., A. Bauer, J. C. Zubriski, J. Bauder. 1977. Crop Curves for Water Balance Irrigation Scheduling in S.E. North Dakota. North Dakota Research Report No. 66. North Dakota Agricultural Experiment Station, Fargo.

11. Stegman, E. C., G. W. Lemert. 1981. Sunflower Yield vs. Water Deficits in Major Growth Periods. Transations of ASAE (in press).

12. Stegman, E. C., D. C. Nelson, 1973. Potato Response to Moisture Regimes. North Dakota Research Report No. 44. North Dakota Agricutlural Experiment Station, Fargo.

13. Stegman, E. C., L. H. Schiele, A. Bauer. 1976. Plant Water Stress Criteria for Irrigation Scheduling. Transactions of ASAE 19(5):850-855.

14. Stegman, E. C., J. D. Valer. 1972. Irrigation Scheduling by Computational Methods; Crop Data and Evaluations in North Dakota. North Dakota Research Report No. 41. North Dakota Agricultural Experiment Station, Fargo.

15. Vitosh, M. L. 1977. Irrigation Scheduling for Field Crops and Vegetables. Extension Bulletin E-1110. Michigan State University, East Lansing.

16. Vitosh, M. L., E. H. Kidder. 1975. Irrigation Management for Corn. Extension Bulletin E-853. Michigan State University, East Lansing.

17. Werner, H. 1978. Irrigation Scheduling - Checkbook Method. M-160. Agricultural Extension Service, University of Minnesota, St. Paul.

18. Werner, H. 1979. Irrigation Checkbook. Irrigators Association of Minnesota. Staples, Minnesota.

IMPLEMENTING A WATER CONSERVATION PROGRAM IN WESTLANDS WATER DISTRICT

WILLIAM R. JOHNSTON, BYRON C. STEINERT, DAVID L. WOOLLEY[1/]
MEMBER
ASAE

Westlands Water District (WWD), a public water agency, covers about 244,000 ha (603,000 acres) on the west side of the San Joaquin Valley of California. The climate is semi-arid with an average annual Class "A" pan evaporation of more than 165 cm (65 inches) and an average annual rainfall of about 18 cm (7 inches) (California, 1975). The soils, for the most part, are composed of gently sloping, recent alluvium fans of marine deposits formed during the middle Tertiary periods (Harradine, 1950). The elevation of the land surface varies from 50 m (165 feet) to 260 m (850 feet) and averages 95 m (312 feet) above sea level.

WWD usually receives 1,418,500 dam^3 (1,150,000 acre-feet) of water annually from the U.S. Bureau of Reclamation (USBR) San Luis Unit of the Central Valley Project. Water is allocated to land within WWD and is delivered on demand to the District's 250 water users through a pressure pipeline distribution system. In addition, about 123,350 dam^3 (100,000 acre-feet) of groundwater are pumped annually by some water users. The total available water supply for the 228,600 irrigible ha (565,000 acres) in WWD is approximately 1,542,000 dam^3 (1,250,000 acre-feet) annually or about 67 cm (2.2 feet). The average annual effective precipitation is only about 3 cm (0.1 foot) and does not contribute significantly to the water supply. WWD's water users adjust their cropping patterns in order to fit anticipated irrigation needs to their available water supply.

It is estimated that eventually about 121,400 ha (300,000 acres) of WWD will need subsurface agricultural drainage. During October, 1980, about 106,400 ha (263,000 acres) of District land had a perched saline water table less than 6.1 m (20 feet) below the ground surface. On 59,500 ha (147,000 acres), the perched water table was less than 3 m (10 feet) below the ground surface and more than 9,700 ha (24,000 acres) had water within 1.5 m (5 feet) of the ground surface. Even though the salinity of the shallow groundwater ranges between 6,000 and 25,000 mg/l total dissolved solids (TDS), the crops grown in the area of shallow groundwater use some of this water to satisfy a portion of their evapotranspiration (ET) demand.

CROPS AND WATER REQUIREMENT

A large variety of crops are grown in WWD. Table 1 lists the average annual ET of the major crops in the District. Column (4) of Table 1 shows the average ET of the crops in centimeters and column (5) shows the portion of the total ET for each crop for a composite of the District. The total average ET of 66.2 cm (26.1 inches) shown in column (5) is only a portion of the water required to satisfy the needs of the crops grown. Additional water is

1/ The authors are: Assistant Manager - Chief of Operations, Senior Engineer, and Water Conservation and Management Specialist, Westlands Water District, Fresno, California.

required for system losses; salinity control, including leaching and reclamation; and for some crops, climate control and germination. When a practical and attainable application efficiency of about 85 percent is considered for the District, the water duty in WWD is about 85 cm (2.8 feet).

Table 1: Evapotranspiration of Crops Grown in Westlands Water District

(1) Crop	(2) Hectares	(3) Percent	(4) Evapotranspiration: Crop	(5) Evapotranspiration: District Composite
			(cm)	(cm)
Alfalfa - Hay	4,100	1.8	110.0	2.0
Alfalfa Seed	7,700	3.4	79.0	2.7
Barley	31,000	13.6	43.0	5.8
Cantaloupes	7,300	3.2	43.0	1.4
Cotton	115,300	50.4	76.0	38.3
Safflower	4,000	1.8	70.0	1.3
Sugar Beets	4,500	2.0	98.0	2.0
Tomatoes	11,300	4.9	64.0	3.1
Wheat	22,500	9.8	49.0	4.8
Fallow	6,700	2.9	--	-
Miscellaneous	17,800	7.8	61.0	4.8
Subtotal	232,200	101.6		66.2
Double Crop	(3,600)	(1.6)		--
TOTAL	228,600	100.0		66.2

INITIAL IRRIGATION SCHEDULING PROGRAM

An initial cooperative irrigation scheduling program was started in 1973 by WWD, the USBR, and a number of District water users.

During this program, the District used a USBR remote terminal in Fresno to access the USBR computer facilities located in the Engineering and Research Center in Denver, Colorado. Potential evapotranspiration (ETP) was determined for each day during the growing season from climate data collected in WWD. The ETP is determined using a modified Penman equation which predicts the ET for a well watered alfalfa crop with 20 cm (8 inches) or more of top growth (Wright and Jensen 1978). The Jensen-Haise formula was initially used to calculate ETP values in WWD. However, it was found that the excessive wind in the area was not taken into account. Because of this problem, new crop curves were developed for use with the modified Penman formula.

Computer models were used to calculate ET for selected crops grown in WWD. A typical crop water use curve for cotton grown during 1980 is shown on Fig. 1.

The quantity of water used by cotton during any given day is determined by multiplying the crop coefficient, kc, obtained from Fig. 1, by the ETP determined with the modified Penman equation.

Soil moisture observations were made at least bimonthly and as often as twice weekly during periods of peak water use to verify the accuracy of the crop water use model.

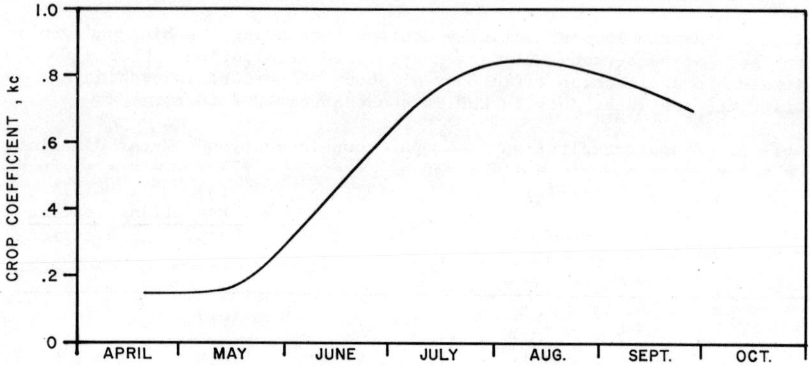

Fig. 1. Crop Curve for Cotton Grown in Westlands Water District During 1980.

This irrigation management service provided participating water users computerized printouts on soil moisture depletion. These printouts and discussions of observed soil moisture depletion assisted the farmers in determining the amount of water to apply and the timing of each irrigation.

One of the major problems of the initial program was the coordination of the calculated ET data to the soil moisture holding capacities of the many different soil types within WWD. This problem was overcome through continuous extensive soil moisture data collection for each major soil type.

The field-by-field irrigation scheduling program was well received. However, WWD could not furnish technical assistance to a large enough portion of the land in the District with the limited personnel available. Because of the limited capabilities of WWD, several water users started their own programs. Others retained the services of commercial irrigation management consultants.

CROP WATER USE GUIDE PROGRAM

In 1978, in order to provide crop water use information to all the water users in WWD with a limited number of personnel, a Crop Water Use Guide was developed. The Crop Water Use Guide has been mailed to all District water users each week since December of that year. Figure 2 is a copy of the June 27, 1980, Guide and instructions on its use.

District climatic measurements and National Weather Service forecasts are used to calculate and forecast crop water use with the modified Penman equation. Crop water use is monitored on selected fields to verify the data calculated by the computerized model and shown on the Guide. The purpose of the Guide is to help the water users estimate the amount of crop water use and schedule the amount and time of the crop irrigations. This approach to irrigation management optimizes water and electrical energy use and helps reduce labor costs while maintaining crop productivity.

Each water user is encouraged to supplement the Guide information with periodic soil moisture sampling or other techniques for monitoring crop water use and to apply only the amount of water required to replenish the root zone and necessary to satisfy salinity control and other cultural needs.

WWD continued to utilize the USBR computer to model crop water use and print the weekly Guide. However, in 1979, WWD installed a remote terminal in its office to access that computer. This eliminated the need for District personnel to travel to the local USBR office to update the Guide.

WESTLANDS WATER DISTRICT - - - CROP WATER USE GUIDE - - - - - - JUNE 27, 1980

①	②	③	④	⑤	⑥	⑦
			----PAST AVERAGE DAILY USE----			-FORECAST-
	APPROX.	TOTAL	PRIOR	PRIOR	LAST	---USE---
	START	USE	20 DAYS	10 DAYS	10 DAYS	10 DAYS
CROP	DATE	TO DATE	518-606	607-616	617-626	627-706
ETP			.31"	.36"	.42"	.37"
ALF. HAY	MAR 1	18.1"	.21"	.25"	.29"	.26"
ALF. SEED	MAR 1	19.0"	.24"	.28"	.33"	.29"
ALMONDS	FEB 20	14.8"	.18"	.21"	.25"	.22"
COTTON	APR 7	6.2"	.07"	.14"	.21"	.23"
COTTON	APR 15	5.5"	.07"	.12"	.18"	.19"
GARLIC	NOV 15	17.0"	.19"	.20"	.17"	.10"
GRAPES	MAR 15	10.5"	.15"	.18"	.21"	.18"
MELONS	APR 15	9.7"	.11"	.19"	.28"	.26"
OLIVES	JAN 1	12.4"	.13"	.15"	.18"	.16"
ONION	NOV 15	18.0"	.19"	.21"	.25"	.22"
PEPPERS	MAR 15	13.4"	.20"	.25"	.29"	.26"
POMEGRANATE	MAR 20	11.2"	.15"	.18"	.21"	.18"
SAFFLOWER	MAR 1	16.3"	.24"	.29"	.34"	.28"
SUGAR BEET	NOV 15	19.9"	.21"	.25"	.29"	.26"
TOMATO-PROC	APR 1	13.4"	.22"	.29"	.34"	.29"
TOMATO-PROC	MAR 1	17.1"	.25"	.29"	.32"	.25"
TOMATO-MKT	MAR 15	17.2"	.25"	.29"	.34"	.29"

GUIDE INSTRUCTIONS

This "Guide" is to assist in determining how much water should be applied by estimating the amount of water the crop has consumed and forecasting water use prior to the next anticipated irrigation date. The columns shown on the "Guide" are defined as follows:

<u>Column</u>:

1. Crops currently being grown in the District. ETP is included in this column.
2. Approximate planting date or date the crop would come out of dormancy. Some crops will show more than one start date.
3. Total crop water use to date is that amount of water in inches a plant has consumed since its start date.

4.5.6. Average daily crop water use in inches for the last 10 days, prior 10 days and prior 20 days. (40-day range)

7. Forecast average daily crop water use in inches for the next 10 days.

ETP. ETP is the evapotranspiration from an extensive surface of a well watered alfalfa crop with 20 cm (8 inches) or more of top growth.

EXAMPLE OF GUIDE USE

The following example shows how to determine the amount of water a cotton crop that was planted on April 15, is forecasted to use through June 30, 1980.

Total water use through the date of the Guide, June 27, 1980 (Column 3)	5.5 inches
3 days at 0.19 inches per day (Column 7)	<u>0.6</u> inches
TOTAL WATER USE	6.1 inches

If an irrigation is to take place on June 30, then 6.1 inches of water will be required to replace the water removed from the plant root zone. If this sample cotton crop was pre-irrigated and the plant root zone was filled, the 6.1 inches of water would probably refill the root zone.

Normally irrigations are not 100 percent efficient, so it is usually necessary to apply an additional 15-20 percent of water. Therefore, it would probably be necessary to use about 7 inches of water as measured at your field delivery or pump for the irrigation in the above example to insure that the root zone is filled.

Fig. 2. Crop Water Use Guide, Guide Instructions and Example of Guide Use

WATER CONSERVATION AND MANAGEMENT PROGRAM

Water conservation and management are extremely important issues in California, not only from the practical, but from the political point of view. Because of the need to utilize a limited water supply to the fullest extent possible and to help the water users in WWD do the best job possible in managing their water supply, the District implemented an expanded Water Conservation and Management Program in 1981. The program was also necessary to show those outside of farming that WWD is doing an outstanding job in managing its limited water supply.

This program involves: (1) expanding the use of the Crop Water Use Guide, (2) development of a District furnished Water Conservation and Management Handbook, (3) development of an on-farm water accounting system, and (4) increasing District furnished water management advise.

The growing season is different in various parts of the District. Therefore, in order to provide more accurate water use data for the crops, the District was divided into three zones, containing approximately 200,000 acres each. Three Crop Water Use Guides are to be developed, one for each zone. In addition, by dividing the District into three zones, the travel time needed by the technicians to monitor crop water use and verify the computer model is reduced.

A Water Conservation and Management Handbook is being developed. The information contained in this Handbook will pertain specifically to the water conservation and management problems of the District's water users. The development of the Handbook will be an ongoing program and inserts will be distributed to District water users as they are developed. A water acoounting system is a part of the Handbook.

In order to plan and monitor the most efficient utilization of a given water supply, it is necessary for the farmer to plan effectively. The District has developed a series of forms which are also part of the Handbook. This will help the water user carry out his planning function. Figures 3 and 4 are forms provided by the District and utilized by the water user. Water users are generally not familiar with metric units; and at this time, it is more important to have the users accept and utilize the information presented than it is to try to convert the farmers to the metric system.

The Farm Water Budget form shown on Fig. 3 is used to assist the water user to plan future cropping patterns and to match crop water needs with his available water supply. This helps the water user determine the quantity of water he should allocate to a specific crop and field. The water user is then encouraged to record the quantity of water applied during each irrigation and determine the balance of his water allocation available for each field on the Field Water Budget form shown on Fig. 4.

Three Water Conservation and Management Specialists are now employed by WWD. A specialist is assigned to each zone of the District. These technicians will assist District water users in evaluating their irrigation systems and practices. The technicians will also provide assistance in determining the soil moisture holding capacities of different soil types and in determining soil salinity problems which would help the water user determine his crop selection and water needs.

In addition, the District is going to install an in-house computer. When the District's computer is installed, all work being done on the USBR computer will be transferred to the District's facilities. This will also enable the District to expand its capabilities. WWD will be able to provide each water user with historical water use and projected crop water use requirements.

WESTLANDS WATER DISTRICT

WATER CONSERVATION AND MANAGEMENT PROGRAM

FARM WATER BUDGET

Farm Name	Water Allocation from Westlands AF
Crop Year	Add Water used for Pre-irrigation Prior to December 31 AF
Irrigated Acreage	Subtract Water to be used to Pre-irrigate Next Year's Crop ( ) AF
Date	Add Groundwater Supply AF
Prepared By:	Farm Water Supply for this Year's Crop AF

CROP	AVERAGE CONSUMPTIVE USE		EFFECTIVE PRECIPITATION	WATER FOR GERMINATION	SALINITY CONTROL	OTHER WATER	CROP WATER REQUIREMENT	APPLICATION EFFICIENCY	FIELD WATER REQUIREMENT	PLANTED AREA	FARM DELIVERY REQUIREMENT
(1)	Inches (2)	AF/Ac (3)	AF/Ac (4)	AF/Ac (6)	AF/Ac (5)	AF/Ac (7)	AF/Ac (8)	% (9)	AF/Ac (10)	Ac (11)	AF (12)
Alfalfa Hay	43.5	3.6									
Alfalfa Seed	31.6	2.6									
Almonds	33.5	2.8									
Barley	16.3	1.4									
Beans (Dry)	13.5	1.1									
Cantaloupes	17.0	1.4									
Corn (Field)	24.2	2.0									
Cotton	29.5	2.5									
Garlic	20.2	1.7									
Grapes	28.3	2.4									
Lettuce (Spring)	10.1	0.8									
Lettuce (Fall)	8.0	0.7									
Olives	31.3	2.6									
Onions	27.3	2.3									
Peppers (Gr)	24.6	2.1									
Pomegranates	31.5	2.6									
Rice	40.3	3.4									
Safflower	27.6	2.3									
Sorghum (Milo)	18.4	1.5									
Sugar Beets	38.9	3.2									
Tomatoes (Proc)	24.9	2.1									
Tomatoes (Market)	24.7	2.1									
Wheat	19.5	1.6									
RESERVE											
TOTAL FARM WATER REQUIREMENT											

Fig. 3 Farm Water Budget Form

WESTLANDS WATER DISTRICT

WATER CONSERVATION AND MANAGEMENT PROGRAM

FIELD WATER BUDGET

FARM NAME				YEAR
FIELD WATER REQUIREMENT	AF/Ac	PLANTED AREA	Ac	DATE PLANTED
ADJUSTMENTS	AF/Ac	WATER BUDGET	AF	DATE HARVESTED
NET WATER REQUIREMENTS	AF/Ac			YIELD

FIELD NO. CROP

METER NUMBER	DATE		METER READING		WATER APPLIED					WATER BUDGET BALANCE
					IRRIGATION			TOTAL		
	ON	OFF	ON	OFF	AF	AF/Ac	INCHES	AF	INCHES	AF
1	2	3	4	5	6	7	8	9	10	11

FIELD WATER REQUIREMENT	AF/Ac	PLANTED AREA	Ac	DATE PLANTED
ADJUSTMENTS	AF/Ac	WATER BUDGET	AF	DATE HARVESTED
NET WATER REQUIREMENTS	AF/Ac			YIELD

FIELD NO. CROP

METER NUMBER	DATE		METER READING		WATER APPLIED					WATER BUDGET BALANCE
					IRRIGATION			TOTAL		
	ON	OFF	ON	OFF	AF	AF/Ac	INCHES	AF	INCHES	AF
1	2	3	4	5	6	7	8	9	10	11

Fig. 4. Field Water Budget Form

CONCLUSION

The WWD program shows that a large agricultural area growing diversified crops can be provided reliable crop water use data designed to assist the water users solve specific water conservation and management problems by a local district.

REFERENCES

1. California, State of, Department of Water Resources (DWR), 1975. Vegetative Water Use in California, 1974. Bulletin No. 113-3.

2. Harradine, Frank F., 1950. Soils of Western Fresno County, California. University of California, College of Agriculture, Agricultural Experiment Station, Berkeley 4, California. 86 p.

3. Wright, James L. and Marvin E. Jensen. 1978. Development and evaluation of evapotranspiration models for irrigation scheduling. TRANS. of the ASAE 21:88-96.

THE IRRIGATION MANAGEMENT PROGRAM:

A NEW APPROACH TO COMPUTER ASSISTED IRRIGATION SCHEDULING

E. Fereres R.E. Goldfien W.O. Pruitt
Member ASAE

D.W. Henderson R.M. Hagan
Asso. Member ASAE

Simple methods of scheduling irrigations using climatological, soil and plant parameters have been available for some time (Chang, 1961; Halkias et al., 1955; Pearl, 1954; Pruitt, 1956; Pruitt, 1964; Stanhill, 1962 and Thompson et al, 1963) and in a few cases they have been widely used (Chang, 1961; Stanhill, 1962 and Thompson et al., 1963). However, precise irrigation scheduling requires an information data base which is difficult for most farmers to obtain. With the advent of high-speed computers, it has become technologically feasible to simulate the processes involved in the transfer of water in the soil-plant-atmosphere system and to use such models for predicting the timing and amounts of irrigations (Jensen, 1969, 1970). Computerized irrigation scheduling services (CISS) based on water budget calculations and field checks are now available in certain areas of the western states either under Federal programs (USDI, 1979) or through private consultants.

Experience in California has shown the while CISS have significant potential for improving on-farm water management practices, general adoption of such services has been limited. Therefore a need exists to provide simplified irrigation scheduling information to the vast majority of farmers who currently irrigate on a calendar basis. Over the last 20 years, significant information has been developed in California on the crop water requirements and on the moisture characteristics of the principal soils (Fereres et al., 1980). That information may now be used to design irrigation scheduling programs which could be extended to the general public. Implementation on a large scale can be achieved only by a program that a large number of growers know about and are willing to try; that is technically sound so that it will be of significant benefit to its users; and that can be made available on a large scale at reasonable cost. Any attempt to meet all of these criteria must involve some compromise, as they can be mutually exclusive to some extent. The Irrigation Management Program (IMP) was designed to meet these criteria insofar as possible (Fereres et al., 1980).

COMPUTER MODEL

The IMP computer model is based on the water budget concept. The allowable depletion (AD) for each day (i) is calculated using:

$$AD_i = RD_i \times AW \times \%\ AD_i \qquad (1)$$

The authors are: E. Fereres, Extension Irrigationist; R.E. Goldfien, Post Graduate Researcher; W.O. Pruitt, Irrigationist; D.W. Henderson and R.M. Hagan, Professors, Department of Land, Air and Water Resources, University of California, Davis.

where RD_i is the effective rooting depth of the crop, AW is the available soil water per unit depth of soil, and % AD is the percentage of the available soil water that the crop can extract without being subjected to water stress. RD is a function of time and is estimated using an assumed relationship between canopy development and root growth in which the crop reaches its maximum rooting depth at about the time it attains full ground cover. The crop coefficient (Kp), relating the crop evapotranspiration (ET_c) to evaporation from a class "A" U.S. Weather Bureau Pan (Epan), has also been well correlated to the percentage of ground cover. RD_i can then be estimated using:

$$RD_i = RD_o + (RD_m - RD_o)\frac{i}{i_m} \qquad (2)$$

where RD_o is the initial rooting depth, assumed to be 4 inches, RD_m is the maximum effective rooting depth, and i_m is the time, in days after planting, that the crop coefficient reaches a maximum. AW may be determined by soil sampling or can be estimated from the soil texture or soil survey information. % AD is a function of several variables including the type of crop, stage of growth, climate and soil moisture retention characteristics. Usually the values of some of these variables are uncertain so a management decision must be made as to what is a "safe" level of depletion. Field trials and experience have established reasonable management practices for many crops and soils, but where uncertainty is large a conservatively low estimate should be used.

The model assumes that the soil profile is full at the start of the season and does not account for rainfall during the growing season as it is usually negligible under Central Valley conditions during the summer months. AD is then calculated on a daily basis using Eq. (1). Next, soil moisture is computed as follows:

$$SM_i = SM_{i-1} - ET_i \qquad (3)$$

where SM_i is the soil moisture above the allowable depletion level on day i, SM_{i-1} is soil moisture on the previous day, and ET_i is the daily ET of the crop calculated from:

$$ET_i = K_{pi} \times Epan_i \qquad (4)$$

where K_{pi} is the crop coefficient and $Epan_i$ is the expected pan evaporation based on historical data. SM is initially set at AD_o and Eq. (3) is used to calculate SM_i. When $SM_i \leq 0$, an irrigation is scheduled for day i with the amount being equal to AD_i. SM_i is reset to AD_i and the cycle continues. In this way a full-season irrigation schedule is simulated using the long-term average data on crop ET.

The year-to-year deviations of ET from average, in California's Central Valley, usually are quite small particularly so during the summer months. It has been found that 90 percent of the time 10-day to 2-week forecasts of ET, based on historical data, are within 10 percent of actual ET (Fereres et al., 1981). Figure 1 shows scheduled irrigation dates, based on actual ET in the respective crop-year, plotted against those based on historical ET. These tests support our hypothesis that a tentative seasonal irrigation schedule based on historical ET data can be quite reliable for planning irrigations and other farm operations.

LEAFLETS

The tentative irrigation schedule and other information are distributed in a one-sheet leaflet. Leaflets are generated by computer and designed to present a large amount of useful information in a way that is versatile and easy to

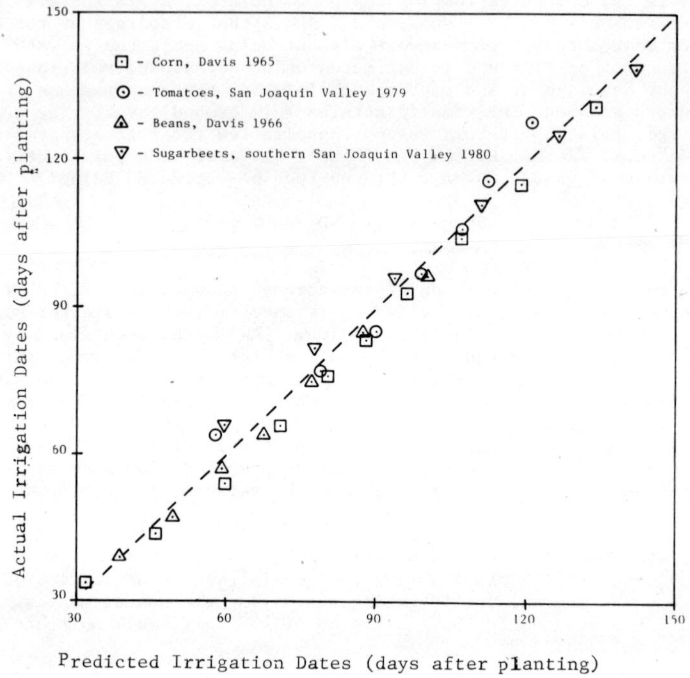

Fig. 1. Comparison of Actual and Predicted Irrigation Dates Based on Actual and Normal ET, Respectively. (After Fereres et al., 1981).

use. Figure 2 is an example of a typical leaflet. The title is used to select the leaflet that is most appropriate for the particular crop, location, planting date, soil type, and allowable depletion. A leaflet can be designed for any situation where these factors are known along with the corresponding Epan and Kp values. This information is available for many crops in several areas of California (SDWR, 1975).

Each leaflet contains the following information for the crop, location, planting date, soil depth, and allowable depletion given by the title: 1) a seasonal cumulative ET curve with lines drawn to indicate the date of irrigation and the amount; 2) a printed irrigation schedule listing the number of days after planting until each irrigation, the cumulative ET up to that time and the net depth of irrigation water; 3) a table giving the expected pan evaporation (Epan), crop coefficient (Kp) and crop ET for one- or two-week intervals throughout the season; 4) a water budget form, on the back of the sheet, that can be used to adjust the IMP to individual needs.

The leaflet can be used to estimate the seasonal irrigation requirements of the crop which can then be used as the basis for scheduling irrigations and other farm operations. The irrigation schedule can be followed as is, or updated as necessary to adjust for variations from expected ET. Because of the relative constancy of the evaporative demand in California's Central Valley from year to year the updated schedule will usually fall very close to the tentative schedule given in the leaflet. Therefore, if a fairly conservative allowable depletion is used, the tentative normal year schedule frequently can be followed without adjustment. Recommended practice, however, is to monitor ET closely and adjust the schedule as necessary.

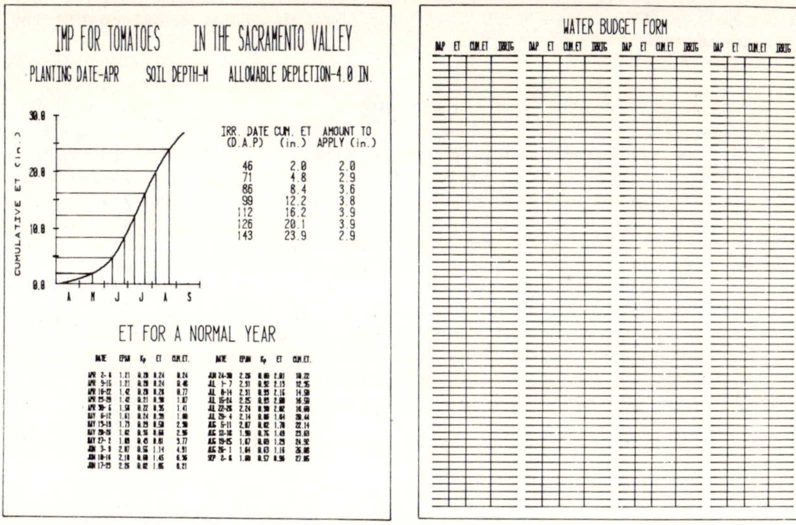

2a. 2b.

Fig. 2 IMP Leaflet for Tomatoes in the Sacramento Valley. Planting Date: April 1; Soil Depth: Medium; Allowable Depletion: 4.0 Inches. a) Front of Leaflet Giving Scheduling Information. b) Water Budget Form on the Back of Each Leaflet.

PROGRAM IMPLEMENTATION

The final phase of the IMP involves the large scale implementation of the program. The ultimate success or failure of the IMP will depend on our ability to make the leaflets widely available and to provide adequate training and assistance in their use. Preliminary discussions involving several state and federal water related agencies have indicated enthusiasm for the program and willingness to cooperate in it. Because the computer program that generates the IMP is fairly simple, it can easily be shared by several agencies and private firms in order to diffuse the cost of the program. This may make it possible to distribute the leaflets at minimal cost to the growers. It is not feasible, however, to attempt to provide leaflets for all possible combinations of crop, location, planting date, soil depth, and allowable depletion as this would involve an enormous number of different leaflets in California alone. Ideally many of the more common leaflets would be readily available with the more unusual situations being handled on demand by designing a new IMP or making adjustments to an existing one. The interagency framework for such distribution has not been worked out at the present time.

Field verification of computer predictions is another important part of the overall program. This is necessary because of uncertainties about the net depth, and uniformity of distribution of water applied at each irrigation. There are also uncertainties in evaluating the crop rooting depth, soil water storage capacity, and allowable depletion. In addition, the spatial variability of soil water holding characteristics within each field, uncertainties in computation of crop ET, particularly in the early growth stages; and the need to account for effective on-farm rainfall make field checks most desirable. Growers need assistance in developing a field check plan and the necessary techniques. Implementation of the field check program must be the responsibility of the grower who can hire personnel to carry out the checks. This service could also be made available through agricultural consulting firms at reasonable cost.

CONCLUSIONS

The IMP offers a compromise between the complexity of more precise methods and the convenience of the user. While its many simplifications may introduce errors, in many cases it is unrealistic to improve the accuracy of irrigation schedules when so much uncertainty exists in irrigation water application and distribution. The IMP's are a simplified practical approach, and we are optimistic that it will receive much greater acceptance than past efforts to involve farmers in the use of technical data to schedule irrigations. IMPs offer potential savings in on-farm water application and reductions in energy and labor costs, as well as potential improvements in yields. Furthermore, they may be used to coordinate and schedule other farm operations. Preliminary workshops to introduce agency personnel to the IMP concept and train them in its use have been held in Fresno and Davis, California. The response to the program was very favorable at both workshops. Smaller local workshops are planned to focus on specific crops and practices within the area.

There is room for improvement of the IMP, especially in the rooting depth sub model and the change in crop coefficients and allowable depletions throughout the season. Further allowances can also be made for using some cultural practices as input data. Future work on the IMP should include such objectives. However, we feel that efforts should now concentrate on disseminating the available information before further refinements are made in the program.

REFERENCES

1. Chang, Jen-hu. 1961 Microclimate of sugarcane. Hawaiian Planters' Record. 56:(3) pp. 195-225.

2. Fereres, Elias, Patricia M. Kitlas, W.O. Pruitt, and R.M. Hagan. 1980. Development of irrigation management programs. Final report, Contract B53142, California State Department of Water Resources, University of California, Davis. 212 pp.

3. Fereres, Elias, Patricia M. Kitlas, Richard E. Goldfien, William O. Pruitt, and Robert M. Hagan. 1981. Simplified but scientific irrigation scheduling. California Agriculture 35:19-21.

4. Halkias, N.A., F.J. Veihmeyer and A.H. Hendrickson. 1955. Determining water needs for crops from climatic data. Hilgardia 24:(4), pp. 207-233.

5. Jensen, M.E. 1969. Scheduling irrigations with computers. J. Soil and Water Conserv. 24(8): 193-195.

6. Jensen, M.E. 1970. Scheduling irrigations using climate-crop-soil data. Amer. Soc. Civil Engr., J. Irrig. and Drain. Div. 96(IRI):25-38.

7. Pearl, R.T. (Ed.). 1954. The calculations of irrigation need. Ministry Agr. and Fish., Techn. Bull. No. 4, H.M.S.O., London.

8. Pruitt, W.O. 1956. Irrigation scheduling guide. Agricultural Engineering 37:3. pp. 180-181.

9. Pruitt, W.O. 1964. Evapotranspiration--a guide to irrigation. California Turfgrass Culture 14:(4), pp. 27-32.

10. Stanhill, G. 1962. The control of field irrigation practice from measure-

ments of evaporation. Israel J. of Agr. Res. 12:(2), pp. 51-62.

11. State of California, Department of Water Resources. 1975. Vegetative water use in California, 1974. Bull. 113-3, 104 p.

12. Thompson, G.D., C.H.O. Pearson and T.G. Cleasby. 1963. The estimation of the water requirements of sugarcane in Natal. Proc. South African Sugar Technologists Assoc. pp. 1-8.

13. U.S. Department of the Interior. 1979. Irrigation management services program, annual report 1979. Water and Power Resource Service*, Engineering and Research Center. Denver, CO.

*Now renamed the Bureau of Reclamation.

TWENTY-FIVE YEARS OF MODERN IRRIGATION SCHEDULING PROMOTIONAL EFFORTS

Marvin N. Shearer
Member ASAE

James Vomocil
Member ASA

Energy conservation and water conservation have a major influence on farmer irrigation practices. But even more important than these, in the farmer's value system, is his net profit. A variety of programs and activities were initiated in 1953 by Oregon State University (OSU) to promote modern irrigation scheduling practices, which is an umbrella term used to describe irrigation scheduling based on physical measurements. These programs have continued to the present time.

The objective of this study was to identify reasons for non-adoption of modern irrigation scheduling by growers in Oregon in spite of a 25 year period of intensive educational and promotional programs. Annual Extension Activity Reports and mass media releases by OSU Extension Service for the period 1953 to 1980 on the subject of irrigation scheduling were reviewed. Activities were evaluated against the diffusion process of technology transfer and apparent reasons for adoption or non-adoption were identified.

It was concluded that behavioral patterns and attitudes of people, not the convenience, accuracy, or reliability of the method used are the major obstacles that must be overcome before widespread adoption of modern irrigation scheduling practices will be achieved.

EARLY EFFORTS

Early irrigation scheduling promotional efforts in Oregon were educational, not service oriented (Shearer 1954). County Extension Agents were trained and given assistance in developing and maintaining programs for which they were responsible and given credit (Shearer 1957). Fifty percent of an irrigation specialist's time was devoted to this activity for a period of 15 years. Since 1968 less intensive leadership has been provided.

The basic program demonstrated a simple technique for scheduling irrigations. Activities were initiated with innovators and legitamizers to identify specific labor saving, pumping cost, and yield benefits. Their testimonials were then used to influence early adopters.

An examination was made of irrigation scheduling techniques that were adapted to farmer use and available in 1953. Insitu measurement of soil moisture depletion with gypsum blocks was selected as the basic tool for estimating soil moisture status and making irrigation recommendations, although tensiometers, evaporation pans, and bank book methods were also used.

The authors are: Marvin N. Shearer, Extension Irrigation Specialist and Professor of Agricultural Engineering, Agricultural Engineering Department, and James Vomocil, Extension Soil Scientist, Soil Science Department, Oregon State University.

Gypsum blocks available in 1953 were difficult to install, insensitive to low moisture tension values and required calibration to soils because electrical current flow was not confined to the blocks. Early research and consultation with manufacturers by Samie and Marsh (1955) and Perrier and Marsh (1958) of the Soils Department at OSU resulted in development of equipment that had reasonably accurate calibrations to soil suction values as low as 0.4 bars, was easy to install, simple to use and inexpensive.

Readings from this equipment were easy to chart and interpret by growers. Soil moisture record charts provided graphic illustrations for growers as they explained their irrigation practices at subject matter meetings. Portland General Electric Company, a local private utility, cooperated in a major way during this period by providing gypsum block reading assistance in 5 counties.

Programs were conducted by Extension Agents in 26 out of 36 counties in Oregon by 1958. Approximately 400 innovative farmers were active in programs and others participated in a cursory way. Continuing support was provided both Extension Agents and farmers through numerous news releases, radio and TV programs, local, regional and national magazine articles and training workshops for gypsum block users.

A drawer full of testimonials from farmers expounding the virtues of modern irrigation scheduling on their farms was collected and publicized. Farmers had 'increased yields,' 'reduced pumping costs,' 'reduced labor costs,' 'completely changed irrigation programs,' and 'reduced the number of irrigations per season.' With such success in just 5 years, it appeared that the strong continuing support provided could be reduced.

Continuing support was reduced with disturbing results. When the gypsum block reading and charting service was no longer provided free and the major support effort was reduced, growers started dropping from the program. Farmers were unwilling to take the responsibility of managing programs themselves or making arrangements for others to do so; other activities had higher priority in their value systems.

IRRIGATION MANAGEMENT SERVICES

Formation of irrigation improvement associations was promoted during the next 8 years to overcome the major problem of providing manpower to make the numerous gypsum block readings required.

These associations were groups of 15 to 30 growers living in close proximity to one another who cooperated in hiring an employee to provide the services desired. Measuring stations in fields were usually read 3 times per week and results were plotted on charts which were placed in growers' pumphouses or at other convenient locations. Periodic meetings were held by the associations to evaluate irrigation programs and the effect of gypsum block readings on their irrigation management decisions.

To remain active, however, even these organizations required someone to serve as a leader and initiate activities to promote the organizations' functions. Individual growers had neither the time nor the interest to promote their association's growth. In their minds, their needs were satisfied and they had other responsibilities of higher priority. They operated the associations very successfully until members retired or sold their farms; then the membership of the organizations reduced in size and became inactive. By 1965 the last association disbanded. Some growers, however, continued modern irrigation scheduling programs on their own.

COOPERATION WITH SPONSORING ORGANIZATIONS

During 1965 and 1966 new programs were developed with fieldmen from a local food processing company. Farmers selling bush beans to the processing company were encouraged to follow gypsum block readings for irrigation scheduling. Forty-two percent of those that did reported increased yields and 5 percent reported better grades. The controlled farms had fewer irrigations, but applied more water per irrigation.

The next year 80 percent of the processing plant growers participated in the program. Results were equally encouraging and most growers benefited from the program. The processing company felt it had demonstrated the value of the program for two years and that farmers should continue on their own -- but they didn't.

Results were similar in the Klamath Falls area where a program was cosponsored by Cooperative Extension and the Klamath Soil Conservation District. Farmers reported they had improved the grade of potatoes, reduced the number of irrigations, increased their yields, and found drainage problems they didn't know existed. But here too, as soon as the outside leadership was withdrawn, other farm activities drew greater priority and the practice ceased.

Since 1967 support and assistance has been given any group or individual interested in providing an irrigation scheduling service. The results have been similar.

In 1967 a soil moisture measuring association was formed in a major vegetable producing county and a retired power company employee acted as manager. He decided to retire in 1971 and the association collapsed.

In 1970 an association was formed by the Extension Service for alfalfa growers in Klamath county. The association lost its gypsum block reader after one year and ceased functioning.

In 1973 a private consultant attempted to offer an irrigation scheduling service for a fee. He underestimated expenses; his reader ran up excessive mileage and caused damage on farms. The program lasted one year.

In 1975 an attempt was made to develop and promote computerized irrigation scheduling in Western Oregon. It was based on pan evaporation data, custom fitted to individual fields. Small weather differences in the area served resulted in farmer skepticism of the system. It is believed, however, that this attitude might be as much an excuse as an explanation.

In 1978 the U.S. Bureau of Reclamation cooperated with the Tualatin Valley Irrigation District in offering computerized irrigation scheduling using a combination of computed evapotranspiration and field measurements with a portable neutron probe (Torvend 1978-80). This program was still in existence during the summer of 1981 but active participants dropped from 27 in 1978 to 7 in 1981 during which period a seasonal charge of $50.00 for each neutron measuring station was in effect (Torvend 1981). As a part of this program, weekly evapotranspiration data was published in two local newspapers. Evidence indicates that this information had been used quite widely by irrigation district members and possibly others. The major cost of this program is still financed by the Irrigation District.

One important lasting benefit occurring from these apparent failures has been the learning that took place by growers during their brief encounter with the programs. It has enabled them to acquire a better understanding of soil moisture characteristics and water requirements of their crops. Most have given evidence of correcting poor irrigation scheduling during spring and fall months which had resulted in over-irrigation and excessive labor and energy costs.

EVALUATION

An evaluation of attempts to promote acceptance of modern irrigation scheduling in Oregon during the last 25 years reveals that not once was a program rejected because of convenience, accuracy, or reliability of techniques used. Results showed that irrigation scheduling was viewed positively by growers and desirable results were documented in terms of economics, energy conservation and water conservation. Analysis tends to indicate behavioral patterns and value judgments of growers may have been the dominant cause for lack of sustained adoption of modern irrigation scheduling.

In their discussion of "Technology Transfer and Human Behavior," Muth and Hendee (1980) point out that the nature of an innovation affects its adoption rate. It must be more beneficial than the tool, method, or idea it replaces and have a relative advantage which is most often judged in economic terms. They also indicate it must be compatible with the farmer's way of doing things and be simple so that he will not be awed by its complexity. It must have try-ability; that is, be of such a nature that it can be tried on a small basis and then expanded as the farmer begins adoption, and the results must be observable so that the effects can be determined and expressed visually.

Beal and Bohlen (1957) found that the adoption process is a complex interrelated series of mental activities rather than a single act. It includes awareness of the practice, interest by the individual to the extent further information is sought, evaluation of the idea of the practice to see whether or not it is viable to the farmer's program, a trial of the program to determine specific gains and costs of the program, and finally adoption of the practice for full-scale use.

The 25 year promotional effort closely followed the requirements of the adoption process. Traditional innovators and early adopters had been selected as cooperators and gave evidence of adoption, but adoption was not sustained when the rules were changed, if only slightly.

Many growers during the early efforts, adopted modern irrigation scheduling when the gypsum block reading service was provided free but rejected it when they were required to perform the field readings themselves. Irrigation management associations were initially accepted, but later rejected when the rules were changed requiring members to continue leadership and promotion of their organizations. The efforts that failed after only one year were sometimes rejected because the nature of the innovations were not acceptable. In most cases, what growers were finally asked to adopt was slightly different from what was initially presented to them.

Problems of this technology transfer are in many ways similar to those accompanying the adoption of fertilizer use and weed control practices. Continual support through mass media is given those practices today by industries, agencies, and educational institutions. This same concentration of support may be required before adoption of modern irrigation scheduling is realized.

CONCLUSIONS

The economic, energy conservation and water conservation advantages of modern scheduling procedures can be demonstrated. Modern irrigation scheduling is compatible with most growers' values and beliefs when provided as a service and can be presented to adopters so that complexity will not prevent understanding. It can be implemented on a trial basis by farmers and the benefits can be rapidly observed and documented. Many growers must have help in properly assessing the beneficial effects and must receive continual support. Successful adoption requires that promoters understand and practice what social scientists have learned about the process of technology transfer as it is affected by human behavior.

REFERENCES

1. Abd-El-Samie, A. G., and A. W. Marsh. 1955. "A tube containing gypsum blocks for following moisture changes in undisturbed soil." Soil Science of Am. Proc. Vol. 18, No. 4, pp. 404-406.

2. Beal, G. M. and J. M. Bohlen. 1957. The diffusion process, Agricultural Extension Spec. Rep. 18, Iowa State College, Ames, IA.

3. Muth, R. M. and J. C. Hendee. 1980. "Technology transfer and human behavior." Journal of Forestry, pp. 141-144.

4. Perrier, E. R. and A. W. Marsh. 1958. "Performance characteristics of various electrical resistance units and gypsum materials," Soil Science, Vol. 86, No. 3.

5. Shearer, M. N. 1954. "A practical aid in teaching soil moisture relationship and irrigation schedules," PNW ASAE Regional Meeting.

6. Shearer, M. N. 1957. "County agents in Oregon are experts in moisture determination," County Agent and Vo-Ag Teacher, Vol. 13, No. 5, pp. 11-12.

7. Torvend, Palmer. Irrigation Management Service Annual Reports 1978-1980, Tualatin Valley Irrigation District, Forest Grove, OR.

8. Torvend, Palmer. 1981. Personal communication.

SUCCESSFUL IMPLEMENTATION OF COMPUTERIZED IRRIGATION SCHEDULING

Ronald J. Brase 1/
Gary L. Jorgensen 2/
Gary D. Jardine 3/

Successful implementation of irrigation scheduling in production agriculture requires that scheduling be integrated into the overall farming operation. Factors which contribute to successful integration are positive grower attitudes, having capable personnel involved, and using scheduling methods appropriate to the farming operation. Since the implementation of scheduling usually requires significant changes in overall farm operations, the grower must be committed to the technology and have confidence in the personnel implementing the concepts.

The purpose of this paper is to present the circumstances which Crop Care Associates has found necessary for successful implementation of irrigation scheduling 4/. From this perspective "success" is defined as a full integration of irrigation scheduling into a farming operation; that is, irrigation scheduling becomes as necessary a tool to the ranch manager as his tractors and implements.

IRRIGATION SCHEDULING

The Scheduling Method Used by Crop Care

At Crop Care, we have chosen to use a computerized evapotranspiration (ET) modeling approach to irrigation scheduling. In our system, the crop evapotranspiration calculation is based on the Penman combination equation (Jensen 1974) with the Pruitt modification (Doorenbos and Pruitt 1977). This approach has proven to be responsive to variations in wind movement and humidity, and is more desirable than other approaches including those based solely on temperature and solar radiation.

Weather data are collected by automated weather stations in our areas of operation. Climatic conditions monitored and used in the Pruitt-modified Penman method include maximum and minimum temperature, maximum and minimum relative humidity, solar radiation, and day/night wind speed.

1/ Principal, Crop Care Associates, Fresno, CA.
2/ Senior Agronomist, Crop Care Associates, Fresno, CA.
3/ Agricultural Engineer, CH2M HILL--Engineers, Planners, Economists, and Scientists, Redding, CA.
4/ CH2M HILL developed the desktop data processing system used by Crop Care Associates to schedule irrigation. This paper discusses one approach for implementing the system which has been successfully used by Crop Care Associates.

Field personnel are responsible for maintaining the accuracy of the various parameters used in the computer model. These include the water-holding capacity of the soil, the type of soil, and crop growth characteristics. Based on data verified by field personnel, the computer model calculates daily crop water use and projects future soil moisture levels.

Why An ET Method?

Our approach to irrigation scheduling is to provide our clients with a service that they can depend on and one that can be closely integrated with all farming operations. A computerized ET method is the best approach for achieving this goal. The characteristics of a model are particularly important in field situations with nonuniform soils, poor internal drainage, salt-affected soils, or where climatic conditions vary from the norm. In newly developed areas where historical data or traditional approaches to irrigation scheduling are lacking, the ET model provides estimates of crop water use to determine irrigation requirements. The model also allows crop water use and soil moisture levels to be integrated into the irrigation scheduling system.

Accuracy, Flexibility, and Adaptability: To provide a viable irrigation scheduling service, the tools used and the methods employed must be accurate. They must also be sufficiently flexible to gather data and process it through the full range of expected working conditions. For instance, if soil moisture tensions are expected to be high, one could not rely solely on tensiometers to measure soil moisture since they break tension above 0.8 to 1 bar. Conversely, gypsum blocks would not be sufficiently accurate below 1 bar but would have a good accuracy above 1 bar of soil moisture tension.

The adaptability of the tools and methods used must be commensurate with the range of expected applications. Instrumentation which requires fixed-point installations (e.g., tensiometers and neutron probe sites) must be removed after harvest to allow for cultivation. Fixed-point instrumentation has its greatest application in perennial crops where installations can be left in place throughout the crop season.

We have found that the most practical method to evaluate soil moisture in row crops is using a soil probe and the "feel method" for determining soil moisture. This method is flexible (not restricted to fixed points) and adaptable to almost any situation. However, the "feel method" for judging soil moisture requires experienced field personnel.

The ultimate system, where practical, is measuring soil moisture with instrumentation and incorporating these data into the ET model. The neutron probe accurately assesses soil moisture, and it has applications on some permanent crops. For annual and other permanent crop situations, we do not feel it offers enough flexibility to meet our needs.

Consideration of Various Factors: It is important to choose tools and methods that can consider the factors which are known to affect plant moisture use. A means of quantifying the effect of these factors on a plant should also be provided. For instance, if a crop is grown in a saline soil environment, plant stress will build more rapidly as the soil dries than it would in a nonsaline environment. If the scheduling tools and methods used cannot account for this fact, they are ignoring an important consideration. A properly designed model for irrigation scheduling can attain a very high degree of accuracy and flexibility, as well as being adaptable to almost any situation.

Predictive Capabilities: Several different methods can be used to determine when to irrigate a crop and how much water to apply. However, the accuracy of their predictive ability varies considerably. Soil moisture measuring techniques (i.e., tensiometers, neutron probes, and the feel method) generally do not have a good predictive ability on their own. They can be an accurate means of determining what the soil moisture level is today; however, their accuracy in projecting what the soil moisture will be tomorrow or in a week is somewhat limited. A common practice is to plot data from indirect soil moisture measurements and extrapolate a curve to a predetermined threshold when an irrigation would be scheduled. In many situations, this is adequate; however, we do not feel that this is a truly predictive approach as it does not adequately consider, among other things, variations in the rate of crop water use, crop canopy development, or unusual soil conditions.

A properly designed ET model can accurately project future soil moisture conditions if proper consideration has been given to such factors as changing climatic conditions, variations in plant leaf canopy, the type of soil, water-holding capabilities of the soil, soil chemistry, and the capabilities of the irrigation system. While not every cropping situation warrants the degree of sophistication included in a full and complete ET model, the predictive accuracy of such a model has proven to be a valuable asset in our commercial applications.

Model Operation and Reporting: Our final reason for utilizing a computerized ET modeling approach is its ease of operation, from the perspective of Crop Care as well as the grower. The computerized approach is quite valuable in freeing personnel from repetitive ET calculations and manual recordkeeping. A computer also allows a more sophisticated approach to modeling by providing a means to easily integrate soil stress coefficients and excess moisture coefficients along with crop coefficients.

Once we made a commitment to computer technology, it became relatively simple to output the results of modeling. Irrigation scheduling reports are generated by the computer for both the grower and the irrigation scheduler. The computer also serves as a recordkeeping system and a "followup reminder" for the irrigation scheduler. The "followup reminder" aspect of our system reminds the scheduler of which fields require inspection and when data were last verified.

CAPABILITIES OF PERSONNEL

Successful implementation of irrigation scheduling at Crop Care Associates is contingent upon the use of competent professionals. Formal education and field experience are necessary tools for a professional irrigation scheduler. We have found that entry-level personnel in irrigation scheduling should possess at least a B.S. degree in plant or soil science, or one of the related engineering disciplines. This type of educational background offers the prospective irrigation scheduler an area of specialty in one of the many disciplines involved in irrigation scheduling. A minimum of 1 year's field experience is usually desirable before an individual becomes responsible for scheduling irrigations and communicating with clients.

A person providing irrigation scheduling services at Crop Care must have a basic understanding of soils, crop growth, hydrology, and computer science. Usually, an individual will specialize in his area of expertise while relying on other staff members to monitor state-of-the-art technology in areas other than his specialty. This individual specialization gives our organization a staff of experts in various fields rather than a single-disciplined staff. An appropriate blend of engineering and agricultural sciences also offers our clients a wide range of capabilities from a single organization.

The primary responsibilities of our irrigation scheduling personnel include the verification of soil moisture levels, observation of crop growth, monitoring cultural requirements, and ongoing client communications. Verification of model data is necessary if the model is to retain its accuracy. One of the most important data items to verify is the calculated soil moisture level. It is through a determination of the current soil moisture level that one ascertains the accuracy of the data used to model the crop. Crop growth must also be monitored to ensure development is limited only by soil moisture conditions and not some other aspects of the plant's environment. Cultural requirements for the crop are also important since their timing usually necessitates the use of a more long-range plan for irrigation scheduling.

We have found that regular communication with clients is imperative if irrigation scheduling is to be properly utilized. This regular communication also allows the scheduler to gain additional information and anticipate potential problems. An irrigation scheduler attains his highest degree of credibility with a client by anticipating problem situations and presenting solutions on a timely basis.

Certain skills are desirable for individuals involved with irrigation scheduling. We have found that the most important, aside from technical ability, is the ability to communicate both in writing and orally. This is important to ensure proper implementation of irrigation schedules as well as being essential for a good client relationship. Another important skill is the ability to integrate various disciplines into a fairly broad-based working knowledge. While certain technical understandings can be acquired, it is important for the irrigation scheduler to be competent when communicating with experienced growers. Recommendations, especially those which go against traditions, must be technically sound and delivered with confidence.

In summary, an individual needs a solid educational background, an ability to communicate, and several years of experience if he is going to stand on his own as a competent irrigation scheduler.

ATTITUDES AND INTEGRATION

The attitude of a client or grower who is utilizing irrigation scheduling technology is of paramount importance to success. The grower must have a commitment to the technology and confidence in the personnel implementing the concepts. This allows scheduling to be smoothly integrated into the farming operation.

We have found that the grower who derives the greatest benefit from our irrigation scheduling service is generally an innovative person who runs a farming operation as a business. In addition to being receptive to new technology, he is always seeking new ways to improve his farming practices. This type of person realizes the importance of integrating new technology into his operation if he is to attain the full potential of his farming operation.

The professional irrigation scheduler who is implementing the technology must not only be capable, but must also have the full confidence of the grower. The irrigation scheduler must be viewed by the grower as a well qualified professional. We have found that the requirements for integrating irrigation scheduling technology into a farming operation are:

(1) begin with sound methods, (2) use appropriate tools for the situation, (3) employ capable personnel, and (4) work with a grower who has the proper attitude.

TYPICAL CLIENT DEVELOPMENT

After irrigation scheduling technology has been implemented for several years, we have noticed that certain trends emerge concerning acceptance of the methodology. The following discussion traces what frequently occurs over a 3-year period when the technology is implemented successfully.

First Year

The first year of implementation consists primarily of gaining a grower's confidence and opening communications. In many instances, the irrigation scheduler is asking the grower to change traditional practices which have been used for many years. The technology is usually applied to only a portion of the farm. It is a critical time since the irrigation scheduler is new to the grower's ranch and usually unknown to the grower himself.

Second Year

Entering the second year, the grower has a much higher level of confidence in the professional scheduler and in the technology. Scheduling will likely be expanded in this year to include 80 to 100 percent of the farm. Field tillage, fertilization, and other cultural operations are timed so they are completed before irrigations are scheduled. Integration into other farm operations proceeds very rapidly.

Third Year

The technology is usually fully implemented by the third year. It includes the full acreage of the farm and is an integral part of the farming operation. It is only at this point that we can say irrigation scheduling has been successfully implemented.

SUMMARY

For a computerized irrigation system to be truly successful, it must be integrated into the total farming operation. Implementation of such a program will require modifications in traditional attitudes and approaches to farming.

The success of the program will not only be related to the adaptability and flexibility of the methods and tools used, but will also be directly related to the grower's commitment to the technology and his confidence in the personnel implementing the program. As a result, the scheduler will need a good educational background, an ability to communicate well with growers, and several years of experience in irrigation scheduling.

REFERENCES

1. Doorenbos, J. and W. O. Pruitt. 1977. Crop water requirements. FAO Irrigation and Drainage Paper No. 24. Food and Agriculture Organization of the United Nations. Rome. 144 p.

2. Jensen, M. E. (ed.) 1974. Consumptive Use of Water and Irrigation Water Requirements. American Society of Civil Engineers, New York, N.Y. 216 p.

3. Jensen, M. E., J. L. Wright and B. J. Pratt. 1971. Estimating Soil Moisture Depletion from Climate, Crop and Soil Data. Trans. of the ASAE 14: 954-959.

4. Kincaid, D. C., D. F. Heermann. 1974. Scheduling Irrigations Using a Programmable Calculator (ARS-NC-12). Agriculture Research Service, U.S. Department of Agriculture. 55 pp.

ON-FARM IRRIGATION SCHEDULING IN SOUTHERN ALBERTA

R. H. McKenzie

D. S. Chanasyk
Assoc. Member ASAE

The basic purpose of irrigation is to supply plants with water as needed to obtain optimum yield and quality of a desired plant constituent (Haise and Hagan, 1967). Efficient irrigation implies complete control of the available soil moisture reservoir. Such control requires knowledge of the soil moisture content at all times. Irrigation scheduling attempts to answer two basic questions posed by irrigators: when to irrigate and how much water to apply. Its advantages include the minimizing of water and energy use as well as deep percolation losses while ensuring that adequate water is available to plants without placing them under stress that might reduce yield. Irrigating within limits prescribed by a scheduling program can also help control salinization.

Various techniques are available for establishing irrigation schedules for a given crop, soil and climatic regime. These include: (1) soil water measurements, (2) plant water indicators and measurement, and (3) meteorological approaches (Haise and Hagan, 1967). These methods, or some combination thereof, have been effectively used in irrigated areas throughout the world. Unfortunately, not all relationships between climate, plants and soil are transferrable from one geographic location to another. As a result, coefficients used in these relationships must be re-evaluated for each specific region that scheduling is attempted.

Basic to the operation of any irrigation schedule is knowledge of the status of soil moisture in a given field throughout the growing season. Such information provides initial moisture contents for all schedules and can provide the ground truth information for meteorologically based schedules. The measurement of soil moisture itself provides the most direct method of scheduling irrigation and has the added advantage of taking irrigation efficiencies into consideration.

The neutron probe offers many advantages as a soil moisture measuring instrument. The main one is the ability to make direct, repetitive, nondestructive measurements of volumetric moisture content at the same site. The amount of water present in the soil profile can then be quite easily calculated and once the water holding capacity of the soil is determined, an irrigation schedule can be established. The neutron probe was used in just such a manner in a program to provide on-farm irrigation scheduling for farmers in the Lethbridge region of Alberta.

HISTORY OF IRRIGATION SCHEDULING IN ALBERTA

The southern part of Alberta has a seasonal consumptive use of 660 mm for

The authors are: R. H. McKenzie, Irrigation Specialist, Alberta Agriculture, Lethbridge, and D. S. Chanasyk, Assistant Professor, Department of Soil Science, University of Alberta, Edmonton.

alfalfa, with rainfall contributing approximately 200 mm to crop moisture use. Early settlers in Alberta realized the need for additional water for crop production. The development of the first irrigation projects in Alberta began in the 1890's. Both federal and provincial governments aided in irrigation development during the late 1940's and early 1950's. Today there are thirteen irrigation districts, which are quasi-municipal bodies operating under the Alberta Irrigation Act. Approximately 400,000 ha of land are now being irrigated, about 80 percent of that area being located in the Oldman River Basin. In 1978, this area represented 3.38 percent of the total crop acreage in Alberta and 8.29 percent of the total value of crop production. Wheat, barley and alfalfa hay are the major irrigated crops in terms of acreage, but other grain and forage crops, oilseeds and specialty crops such as vegetables and sugar beets are also grown (Alberta Agriculture, 1978).

Until the 1950's, irrigation water was applied to fields mainly by surface methods and irrigation was normally scheduled on a calendar basis. In the 1950's better designed distribution systems and land-levelling gave irrigators improved control of water being applied. Sprinkler irrigation became common in the 1960's giving irrigators very good system control. These factors lead to improved water management and irrigators began to realize that it was important to know when to irrigate and how much water to apply. Methods used to schedule irrigation included the calendar method, the stage of crop development and the "feel" method of determining soil moisture. In 1966, Alberta Agriculture, in cooperation with the Research Branch of the Canada Department of Agriculture, began offering a program called the Irrigation Gauge. This program was basically a soil moisture budget. It involved measuring the amount of soil moisture at the beginning of the growing season and rainfall and evaporation on a daily basis throughout the growing season at various locations within the irrigation district. With the initial soil moisture at the top of the ledger, evapotranspiration was entered as a debit, decrementing soil moisture. The difference between field capacity and the calculated soil moisture value was the soil moisture deficit or the irrigation requirement. Following irrigation, the amount of water applied was entered as a credit, as was rainfall whenever it occurred. Information on evapotranspiration and rainfall were provided to farmers through radio and local newspapers. Bulletin boards with weekly data were also placed near post offices in most of the communities. Farmers, however, were responsible for keeping their own records of soil moisture. Initially, the program was well accepted by farmers, but the popularity of the program soon declined. One major disadvantage of the gauge method was that the efficiency of the irrigation system was not taken into account. The result was that incorrect amounts of applied irrigation water were frequently entered into the ledgers causing errors in predicting the timing and the amount of the next irrigation. Since these errors were cumulative, the moisture status of the soil could be grossly in error by the end of the season. As well the program provided only general recommendations for a locality rather than specific recommendations for individual fields (Hobbs and Krogman, 1968). The program did, however, serve to educate farmers that crop appearance often did not indicate a need for irrigation until after yield reduction had occurred.

In 1978, the Irrigation Gauge program was modified. Instead of taking only one soil moisture sample at the beginning of the season, soil moisture samples were taken weekly from each field. Soil moisture contents were calculated and recommendations on irrigation requirements were mailed to the farmers. This method proved to be far more successful. However, problems arose with this method of scheduling as well. Obtaining and weighing soil samples, calculating soil moisture levels and mailing out irrigation recommendations proved very time consuming. Time delays in getting the information back to the farmer often resulted in outdated recommendations.

During the past several years, high peak demands for water, particularly in July and August, have caused water shortages in some irrigation districts.

These shortages have resulted in an increased interest in irrigation scheduling and created a public demand for an improved scheduling program.

NEUTRON PROBE METHOD

In 1980, in order to improve the Irrigation Gauge Program, a new scheduling program, which involved monitoring soil moisture in fields on a regular basis using a neutron probe, was offered to farmers. All seven of the Alberta Agriculture Irrigation Specialist Offices were equipped with a neutron probe. Each specialist was responsible for providing the irrigation scheduling service to interested farmers in his respective area (see Fig. 1). The use of the neutron probe by the Lethbridge Irrigation Specialist to provide on-farm irrigation scheduling for farmers in the Lethbridge region will be discussed and its suitability for the program assessed.

Scheduling Details

The new program was offered on a limited basis in the first year, 1980. Seven of the ninety irrigation farmers on the Irrigation Gauge Program were placed on the new irrigation scheduling program using the neutron probe. Twenty seven fields, ranging in size from 20 ha to 64 ha, were involved in the program. Crops irrigated included mainly sugar beets, soft wheat, barley and alfalfa. Soil textures were clay loam or sandy clay loam and all fields were sprinkle irrigated, seven using center pivot systems. Normally only one monitoring site, chosen with the farmer, was located within each field. Field capacity values for each field were determined several days after a 100 mm rain in May and wilting point moisture contents were determined using pressure plate apparatus. Rooting depths were adjusted according to the type of crop and stage of growth. The irrigation specialist made recommendations for each field based on weekly moisture measurements. The frequency and timing of the moisture measurements were determined by the specialist. Generally, irrigation was recommended when the soil water reservoir reached fifty percent of capacity. Each farmer was given the recommendation upon completion of the measurements. Figure 2 shows the scheduling sheet used in the program. Some farmers wished to be told when to irrigate and how much water to apply while others wished only information on how much water to apply. A soil moisture depletion curve for the surface zone of one of the irrigated fields is shown in Fig. 3.

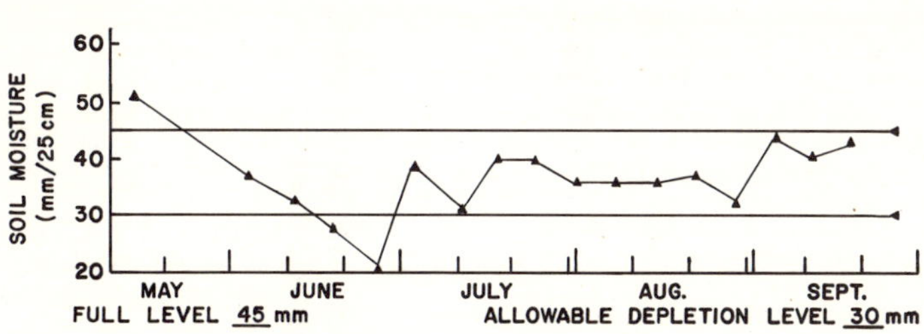

FIG.3 MOISTURE DEPLETION CURVE FOR 10-35 cm DEPTH

At the present time, the irrigation scheduling service is offered free of charge to the irrigation farmers. In 1981, the number of fields for which the scheduling service has been requested from the Lethbridge office has risen to 190. A nominal fee for the service might be instituted if the number of farmers requesting the service from each office increases.

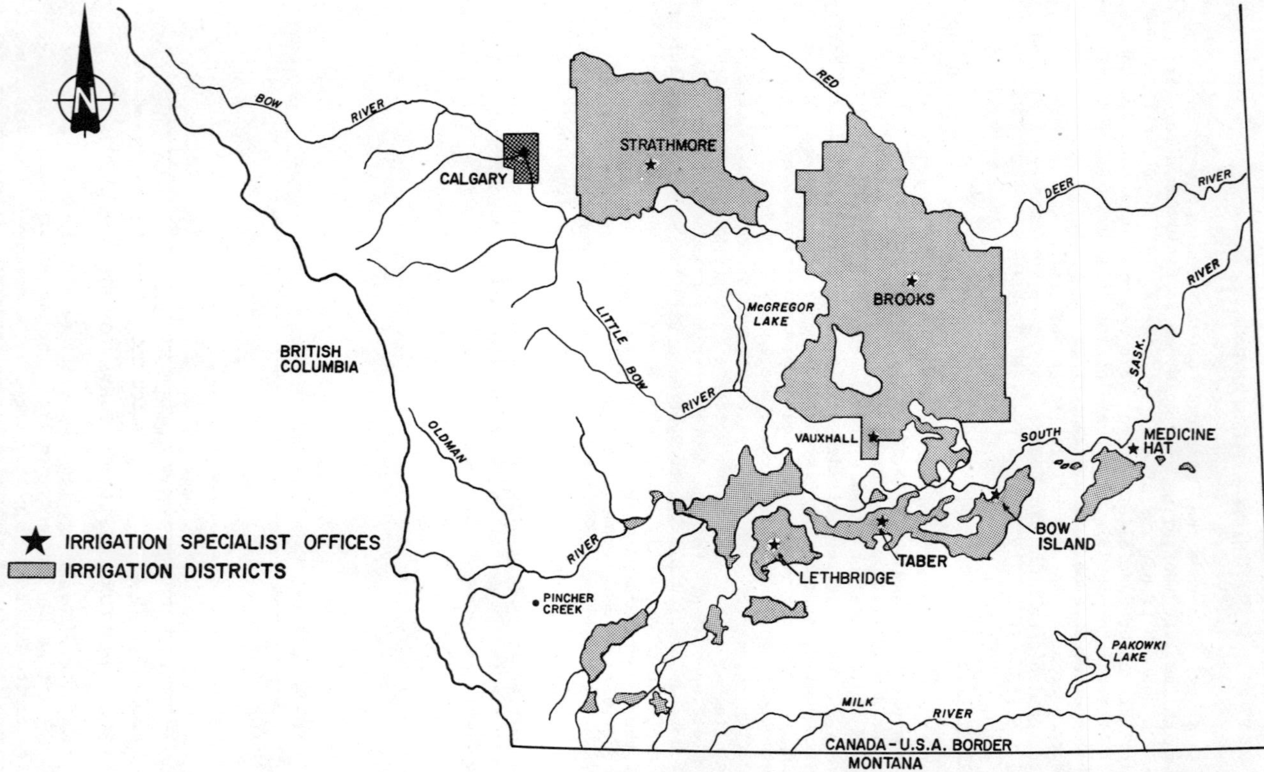

FIG. I IRRIGATION DISTRICTS AND IRRIGATION SPECIALIST OFFICES IN SOUTHERN ALBERTA

IRRIGATION SCHEDULING

Farmer: ______________________ Date: ______________

Field: ______________________ Crop: ______________

Soil Moisture Information

DEPTH (cm)	FIELD CAPACITY mm.	SAFE MOISTURE DEPLETION mm.	PRESENT MOISTURE mm.	ACTUAL MOISTURE DEPLETION mm.
10–35				
35–60				
60–85				
85–110				
TOTAL				

In a ______________ cm. Root Zone:

The PRESENT MOISTURE DEPLETION is __________ mm.

The SAFE MOISTURE DEPLETION is __________ mm.

Available Moisture Remaining is __________ mm.

Daily Crop Moisture Use is __________ mm.

Irrigation Recommendations:

____

____

____

____

____

For Further Information Contact:
Irrigation Specialist Office
Lethbridge, Alta. Phone 329-5137

Alberta
AGRICULTURE

FIG. 2 IRRIGATION SCHEDULING CARD DELIVERED TO FARMER

CONCLUSIONS

All seven farmers on the program found the scheduling information beneficial in terms of planning irrigations for each field, planning overall farm operations and making farm decisions. Some of the farmers realized that they had previously been under-irrigating their crops. All seven have requested to be included in the program for 1981. Based on their interest, the program is considered a success and appears to overcome problems encountered by earlier, unsuccessful programs. Ever increasing water and energy costs, and concern over salinization can be expected to create even greater demand for the program in the future.

FUTURE PLANS

The ultimate goal in the irrigation program is to combine regular soil moisture measurements with the meteorological prediction of crop moisture use, using computer facilities to predict irrigation requirements. It is hoped that a computer program presently being developed by the Research Branch of the Canada Department of Agriculture at Lethbridge can be coupled with the irrigation scheduling program using the neutron probe. With the aid of the computer program, an irrigation specialist could predict when irrigation is required, thereby reducing the number of visits he would have to make to a field during a season. Occasional neutron probe readings would provide a quick, yet reliable means of checking the accuracy of the predictions. The probe would also be used to provide initial moisture contents needed for the program.

REFERENCES

1. Alberta Agriculture. 1978. Agriculture Statistics Yearbook.

2. Haise, H.R. and R.M. Hagan. 1967. Soil, plant, and evaporative measurements as criteria for scheduling irrigation. In: Hagan, R.M., H.R. Haise, and T.W. Edminster. Eds. Irrigation of agricultural lands. Agronomy #11. American Society of Agronomy. pp. 577-604.

3. Hobbs, H. and K.K. Krogman. 1968. An example of research-extension-grower co-operation. Agricultural Institute Review. January-February issue. pp. 23-24.

SUMMARY AND CHALLENGES

Marvin E. Jensen[1]
Member ASAE

I appreciate having the opportunity to participate in this symposium and to preview the papers that were presented. I have enjoyed this assignment because it involves one of my favorite subjects in technology development.

It has been gratifying to see the increased attention that irrigation scheduling technology has received during the past decade. Although many ideas and experiences presented are not new, some new approaches were proposed. Equally important has been our discussion of traditional scheduling concepts that have been unacceptable for continued on-farm use. Identifying non-acceptable methodology should provide valuable guidelines to those engaged in research, technology development, and teaching. We have replowed a lot of old ground during the past three decades. I hope that this symposium has clearly shown that approaches which have failed to gain acceptance during the past three decades may be even more subject to failure in the future because more viable alternatives are now available. Also, we must learn to transfer basic principles and experiences and stop using limited resources to reinvent the wheel.

Some new approaches and evaluations presented are existing. For example, crop canopy temperature differences are related to plant water stress, particularly in arid climates. The relationships may be complex, but they have potential applications in enabling rapid monitoring of the soil water status in fields. A procedure was presented for estimating changes in evaporation rates from soil with time following partial wetting of the soil surface. Procedures were presented for scheduling irrigations to control the leaching fraction using a portable four-electrode salinity probe. A strategy was presented for evaluating alternative scheduling techniques for a large volume, traveling sprinkler system in an area where rainfall provides much of the water needed. I was pleased to see a renewed interest in using historical or expected evapotranspiration rates combined with soil characteristics to approximate irrigation times for the season. By using computers these curves and expected schedules now can be generated very economically for individual farms. They can be very effective in arid areas, but they also can be very informative to irrigators in semiarid and subhumid areas. I consider this type of information to be a necessary part of a service provided by consultants to their clients.

The primary justification for expenditure of public funds for research and development is the development of improved technology for use by farmers or by service groups in helping farmers achieve their management objectives. The resulting benefits might be measurable in terms of increased net returns, reduced semitechnical or labor input, increased agricultural

[1] The author is M. E. Jensen, National Research Program Leader, Water Management, National Program Staff, Agricultural Research Service, USDA, Beltsville, MD 20705.

production from a limited resource (water), or they may simply result in farmer peace of mind in knowing that the best decisions have been made within existing constraints. Although the ultimate user of new technology is the farmer, and because all papers were volunteered, no farmers appeared on this program. Unfortunately, this is not unusual. Research and development is often carried out before the probable acceptance of a new product is assessed. We need to reverse this wasteful trend. I hope that this symposium and these proceedings will significantly affect future research and development of technology for scheduling irrigations by concurrent appraisal of probable acceptance by users.

Both private and agency irrigation scheduling service groups need to be concerned about the probable acceptance of any new or improved technology under development. Survival of private service groups depends on usable and cost effective technology that provides both preseason and real time information that farmers want and need to manage irrigations. Agencies also must realistically consider the required acceptance rate of a new technology to justify continuing research and development efforts. There are over 300,000 irrigated farms in the United States, most of which are in the West. Is an acceptance rate of only one percent per year or about 30,000 users in 10 years unreasonably low? How many farmers currently use technologies that have been promoted and taught in schools for the past 10, 20, or 30 years?

Finally, we need to remember that most farmers and irrigators do a pretty fair job of scheduling their irrigations without the use of instruments and with very limited decision-making data. They will continue this practice unless the technology we provide results in directly measurable or perceived benefits with little or no additional costs or demands on their time. If benefits are not apparent, the probability of acceptance and continued use of new technology may be very small unless there are indirect benefits to the general public and the service is subsidized.

DEFINITIONS

Each author apparently interpreted the symposium independently--no specific topics were assigned. As a result, not all aspects of scheduling were covered.

The following definitions of irrigation scheduling, and its scope, and scheduling services are intended to make my summary and challenges more meaningful.

Irrigation scheduling is a planning and decision-making activity that the farm manager or operator of an irrigated farm is involved in before and during most of the growing season for each crop that is grown. The required decision-making data include: (1) the current level and expected change in available soil water for each field over the next 5 to 10 days; (2) current estimates of the probable latest date of the next irrigation on each field to avoid adverse effects of plant water stress, and the earliest date that the next irrigation can be given to permit efficient irrigation with the existing system (For high frequency, manually operated irrigation systems, this involves the irrigation rate or cycles per day); (3) the amount of water that should be applied to each field, if the irrigator is able to control or measure that amount, which will achieve a high irrigation efficiency and the targeted soil water level; and (4) some indication of the adverse effects of irrigating a few days early or late, of applying too little or too much water, or perhaps terminating irrigations for the season. These adverse effects would be mainly on the marketable value of the crop involved (Jensen 1975). In addition, irrigation scheduling may require consideration of irrigation costs, including electrical load management to limit demand charges, salinity control, optimizing soil moisture for harvesting root crops, providing adequate soil moisture for planting the next crop, and maximizing the potential soil water storage capacity to retain seasonal and overwinter precipitation.

Another important basic concept of irrigation scheduling with most irrigation systems is that the soil water reservoir, and not plant water stress, is being managed. Lead time is always important. As a general rule, the farmer would like to know the tentative date of the next irrigation, or possibly of the next two irrigations, immediately after an irrigation. The amount of water to be applied is often a secondary consideration because the time schedule can be modified to fit that amount using the existing system.

An irrigation scheduling service is a modern service, based on the latest irrigation science and technology, which provides up-to-date information on the status of available water in individual fields; projected date of next irrigation, if another will be needed, or daily rate water should be applied with high frequency irrigation systems to maintain the desired soil water level in each field. The service recommends the allocation and time of water application when either the water supply or the application cost is the primary variable input that will control the net return from the farming enterprise during the current cropping season. The service also recommends the allocation of water and the timing of applications to optimize net returns when another variable like fertilizer, will limit net returns. It provides related recommendations concerning the operation of the farm irrigation system to improve the uniformity of water distribution, reduce water losses, maintain a favorable salt balance in the soil, etc., to improve the managerial skills of the operator or manager and increase net returns from the farming enterprise (Jensen 1975).

SCOPE OF PAPERS

Many papers presented at this symposium and included in these proceedings concentrate on certain components of irrigation scheduling. A few papers summarized experiences encountered with various scheduling methods. Others summarized evaluations of various scheduling techniques. Table 1 highlights the major areas covered by the various authors. It provides readers with a guide for seeking information on a specific issue. The listing of components of practical irrigation scheduling technology also should provide guidelines to those involved in research, development, and teaching.

GAPS

Some of the subjects not covered or that were not covered adequately are:

1. Estimating the amount of water used in evapotranspiration from the capillary fringe associated with shallow water tables (Item 2B(4) in Table 1).

2. Adjusting the schedules to fit the irrigation system capacity and the cultural operations that must be performed (cultivation, spraying, cutting, and baling forage crops, etc.)(Item 2E).

3. Estimating irrigation amounts (Items 2D and 3A(4)).

4. Communications problems involved in delivering information to farmers and obtaining feedback data on recent events affecting future irrigations (Item 2J(3)).

5. Evaluating real time optimization of irrigations (Item 3A(6)).

6. Technology transfer mechanisms including teaching and training workshops (Item 4B).

CHALLENGES

- o Establish several basic technologies that can be taught uniformly in different regions. Some technologies proposed in these proceedings reflect strong personal biases that could result in discontinuities in education and training programs if personnel were to change within educational and research institutions. An example of a basic technology that could be taught in all regions is the use of mean expected evapotranspiration rates for each crop planted at a specific time. This method works very well in arid climates. Results also show that for deep-rooted crops it can work quite well in semiarid climates, especially if some simple adjustment is made to account for major deviations in climate.

- o Determine the acceptability and practicality of requiring farmers to make certain measurements, for example by installing soil instruments and keeping soil moisture records before significant resources are invested into developing a "farmer-oriented" scheduling technology. (See the paper by Shearer and Vomocil for an excellent statement on technology transfer and human behavior).

- o Update farmer-oriented technological concepts taught in agronomy, soils, and agricultural engineering by phasing out emphasis on scheduling techniques and programs that have a very short life or are totally rejected when growers are required to install, maintain, and regularly read instruments or keep detailed records.

- o Critically evaluate and develop guidelines for training new employees seeking employment with service agencies and consultants. Most of the key individuals in successful private service organizations spent one to five years working with experienced individuals or groups after completing a bachelor or masters degree in agriculture or agricultural engineering. Irrigation scheduling service organizations generally are not able to hire new trained employees, but must train their own.

- o Development of new or improved irrigation scheduling technology must be complete to be implemented. Developers should consider costs and related operational issues, equipment maintenance and skill requirements, minimum levels or standards of accuracy of required measurements, estimates and predictions, and related user requirements. Reliability is also very important as frequent instrument failure or only one 5- to 10-day gap in information during the growing season can cause complete loss of confidence in a technology.

REFERENCES

1. Jensen, M. E. 1975. Scientific irrigation scheduling for salinity control of irrigation return flow. Environmental Protection Technology Series EPA-600/1-75-064, 92 p.

Table 1. Irrigation Scheduling Technology Subjects and Symposium Papers That Address These Subjects

Principal subject area	Papers authored by
1. Irrigation scheduling strategies, techniques, and concepts	Hammond et al.; Boggess et al.; Schoney et al.; Rhoades et al.; Slack et al.; Lambert et al.; Pinter and Reginato; Burt and Lord.
2. Components of irrigation scheduling technology	
A. Assessing available soil water	
(1) Direct measurement (gravimetric, neutron probe and remote sensing methods) including problems, required skills and costs	McKenzie and Chanasyk
(2) Indirect measurement (tensiometers, soil moisture blocks, salinity sensors and soil probes)	Cary; Phene et al.; Rhoades et al.
(3) Indirect by measuring plant characteristics (leaf temperature, plant water stress, etc.)	Slack et al.; Pinter and Reginato
(4) Relative evapotranspiration	Reicosky
(5) Remote sensing of soil and plant surfaces and use of models	Hatfield; Pinter and Reginato
B. Estimating soil water depletion rates and control of soil salinity	
(1) Rate of change in soil water as indicated by measurement and sensors	Cary
(2) Evaporation rate from pans or other evaporimeters	Lambert et al.; Westesen and Hansen
(3) Meteorological data and plant growth models	Wright; Lambert et al.; Harrington and Heermann; Lundstrom et al.
(4) Adjusting for shallow water table effects	None
(5) Monitoring salinity levels	Rhoades et al.

Table 1. Irrigation Scheduling Technology Subjects and Symposium Papers That Address These Subjects

Principal subject area		Papers authored by
C.	Estimating irrigation dates	
	(1) Use of measured or estimated soil water depletion rates or crop stress and targeted depletion levels	Hatfield; Cary
	(2) Procedure 2C(1) plus estimated expected precipitation	Lambert et al.
	(3) Procedure 2C(1) or 2C(2) plus expected changes in depletion rates due to expected changes in climate and plant growth	Harrington and Heermann
D.	Estimating amounts of water to apply	Lambert et al.
E.	Adapting schedules and amounts to the capacity of the irrigation system or available flow rate	Hammond et al.; Schoney et al.
F.	Irrigation canal capacity and modes of operation to meet on-farm water delivery needs	Burt and Lord
G.	Optimizing irrigation dates and amounts to achieve management objectives (physical and economic aspects)	Boggess et al.; Schoney et al.; English et al; Stetson
H.	Farmer-oriented tools and instrumentation	Slack et al.; Westesen and Hansen; Lundstrom et al.; Shearer and Vomocil
I.	Controlling high frequency and automated irrigation systems	Phene et al.; Rhoades et al.; Boesch et al.
J.	Irrigation scheduling services	
	(1) General information on ET rates by crop	Fereres et al.; Johnston et al.
	(2) Field-by-field services	Lyford and Schild
	(3) Communications and monitoring technologies	Lyford and Schild

Table 1. Irrigation Scheduling Technology Subjects and Symposium Papers That Address These Subjects

Principal subject area	Papers authored by
(4) Computers and related hardware and software systems	Crouch et al.; Lyford and Schild; Johnston et al.; Harrington and Heermann; Shaffer et al.
3. Evaluations based on real data and experience	
A. Components of irrigation scheduling technologies	
(1) Soil water assessment by direct and indirect methods	Hatfield; Rhoades et al.
(2) Soil water depletion rates	Cary
(3) Irrigation dates	Cary, Fischbach
(4) Irrigation amounts	None
(5) Irrigation schedules and scheduling techniques	Hammond et al.; Slack et al.; Lambert et al.; Johnston et al.
(6) Optimization on a real time basis	None
(7) Scheduling methods and services	Brase et al.; Johnston et al.; Shearer and Vomocil
B. Dependability and reliability of technologies	Westesen and Hansen; Lundstrom et al.
C. Skills required and costs	
(1) Technical skills required	Westesen and Hansen; Lundstrom et al.; Brase et al.
(2) Costs	Lyford and Schild
(3) Communications	Brase et al.
(4) Grower or farmer acceptance	Brase et al.
4. New concepts and approaches	
A. Technologies	Hatfield; Fereres et al.; Pinter and Reginato
B. Technology transfer including teaching and training workshops	None